Reteach Workbook

TEACHER'S EDITION
Grade 6

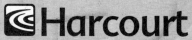

Orlando • Boston • Dallas • Chicago • San Diego
www.harcourtschool.com

Copyright © by Harcourt, Inc.

All rights reserved. No part of this publication may be reproduced or transmitted in any form or by any means, electronic or mechanical, including photocopy, recording, or any information storage and retrieval system, without permission in writing from the publisher.

Permission is hereby granted to individual teachers using the corresponding student's textbook or kit as the major vehicle for regular classroom instruction to photocopy complete student pages from this publication in classroom quantities for instructional use and not for resale.

Duplication of this work other than by individual classroom teachers under the conditions specified above requires a license. To order a license to duplicate this work in greater than classroom quantities, contact Customer Service, Harcourt, Inc., 6277 Sea Harbor Drive, Orlando, Florida 32887-6777. Telephone: 1-800-225-5425. Fax: 1-800-874-6418 or 407-352-3445.

HARCOURT and the Harcourt Logo are trademarks of Harcourt, Inc.

Printed in the United States of America

ISBN 0-15-320818-X

1 2 3 4 5 6 7 8 9 10   022   2004 2003 2002 2001

CONTENTS

▶ Unit 1: NUMBER SENSE AND OPERATIONS

▶ **Chapter 1: Whole Number Applications**
1.1 Estimate with Whole Numbers 1
1.2 Use Addition and Subtraction 2
1.3 Use Multiplication and Division ... 3
1.4 Problem Solving Strategy:
     Predict and Test 4
1.5 Algebra: Use Expressions 5
1.6 Algebra: Mental Math
     and Equations 6

▶ **Chapter 2: Operation Sense**
2.1 Mental Math: Use the Properties .. 7
2.2 Algebra: Exponents 8
2.4 Algebra: Order of Operations 9
2.5 Problem Solving Skill: Sequence
     and Prioritize Information 10

▶ **Chapter 3: Decimal Concepts**
3.1 Represent, Compare, and
     Order Decimals 11
3.2 Problem Solving Strategy:
     Make a Table 12
3.3 Estimate with Decimals 13
3.4 Decimals and Percents 14

▶ **Chapter 4: Decimal Operations**
4.1 Add and Subtract Decimals 15
4.2 Multiply Decimals 16
4.4 Divide with Decimals 17
4.5 Problem Solving Skill: Interpret
     the Remainder 18
4.6 Algebra: Decimal Expressions
     and Equations 19

▶ Unit 2: STATISTICS AND GRAPHING

▶ **Chapter 5: Collect and Organize Data**
5.1 Samples 20
5.2 Bias in Surveys 21
5.3 Problem Solving Strategy:
     Make a Table 22
5.4 Frequency Tables and
     Line Plots 23
5.5 Measures of Central Tendency 24
5.6 Outliers and Additional Data 25
5.7 Data and Conclusions 26

▶ **Chapter 6: Graph Data**
6.1 Make and Analyze Graphs 27
6.2 Find Unknown Values. 28
6.3 Stem-and-Leaf Plots
     and Histograms 29
6.5 Box-and-Whisker Graphs 30
6.6 Analyze Graphs 31

▶ Unit 3: FRACTION CONCEPTS AND OPERATIONS

▶ **Chapter 7: Number Theory**
7.1 Divisibility 32
7.2 Prime Factorization. 33
7.3 Least Common Multiple and
     Greatest Common Factor 34
7.4 Problem Solving Strategy:
     Make an Organized List 35

▶ **Chapter 8: Fraction Concepts**
8.1 Equivalent Fractions and
     Simplest Form 36
8.2 Mixed Numbers and Fractions 37

8.3 Compare and Order Fractions 38
  8.5 Fractions, Decimals, and
      Percents 39

Chapter 9: Add and Subtract Fractions and Mixed Numbers
  9.1 Estimate Sums and Differences ... 40
  9.3 Add and Subtract Fractions 41
  9.4 Add and Subtract Mixed
      Numbers 42
  9.6 Subtract Mixed Numbers 43
  9.7 Problem Solving Strategy: Draw
      a Diagram 44

Chapter 10: Multiply and Divide Fractions and Mixed Numbers
  10.1 Estimate Products and
       Quotients 45
  10.2 Multiply Fractions 46
  10.3 Multiply Mixed Numbers 47
  10.5 Divide Fractions and
       Mixed Numbers 48
  10.6 Problem Solving Skill: Choose
       the Operation 49
  10.7 Algebra: Fraction Expressions
       and Equations 50

Unit 4: ALGEBRA: INTEGERS

Chapter 11: Algebra: Number Relationships
  11.1 Understand Integers 51
  11.2 Rational Numbers 52
  11.3 Compare and Order Rational
       Numbers 53
  11.4 Problem Solving Strategy: Use
       Logical Reasoning 54

Chapter 12: Algebra: Operations with Integers
  12.2 Add Integers 55
  12.4 Subtract Integers 56
  12.5 Multiply and Divide Integers 57
  12.6 Explore Operations with
       Rational Numbers 58

Unit 5: ALGEBRA: EXPRESSIONS AND EQUATIONS

Chapter 13: Algebra: Expressions
  13.1 Write Expressions 59
  13.2 Evaluate Expressions 60
  13.4 Expressions with Squares and
       Square Roots 61

Chapter 14: Algebra: Addition and Subtraction Equations
  14.1 Connect Words and
       Equations 62
  14.3 Solve Addition Equations 63
  14.4 Solve Subtraction Equations 64

Chapter 15: Algebra: Multiplication and Division Equations
  15.2 Solve Multiplication and Division
       Equations 65
  15.3 Use Formulas 66
  15.5 Problem Solving Strategy:
       Work Backward 67

Unit 6: GEOMETRY AND SPATIAL REASONING

Chapter 16: Geometric Figures
  16.1 Points, Lines, and Planes 68
  16.3 Angle Relationships 69
  16.4 Classify Lines 70

Chapter 17: Plane Figures
  17.1 Triangles 71
  17.2 Problem Solving Strategy:
       Find a Pattern 72

17.3 Quadrilaterals **73**
17.4 Draw Two-Dimensional
Figures **74**
17.5 Circles **75**

▶ **Chapter 18: Solid Figures**
18.1 Types of Solid Figures **76**
18.2 Different Views of Solid
Figures **77**
18.4 Problem Solving Strategy:
Solve a Simpler Problem **78**

▶ **Chapter 19: Congruence and Similarity**
19.1 Construct Congruent Segments
and Angles **79**
19.2 Bisect Line Segments
and Angles **80**
19.4 Similar and Congruent Figures **81**

▶ **Unit 7: RATIO, PROPORTION, PERCENT, AND PROBABILITY**

▶ **Chapter 20: Ratio and Proportion**
20.1 Ratios and Rates **82**
20.3 Problem Solving Strategy: Write
an Equation **83**
20.4 Algebra: Ratios and
Similar Figures **84**
20.5 Algebra: Proportions and
Similar Figures **85**
20.6 Algebra: Scale Drawings **86**
20.7 Algebra: Maps **87**

▶ **Chapter 21: Percent and Change**
21.1 Percent **88**
21.2 Percents, Decimals, and
Fractions **89**

21.3 Estimate and Find Percent
of a Number **90**
21.5 Discount and Sales Tax **91**
21.6 Simple Interest **92**

▶ **Chapter 22: Probability of Simple Events**
22.1 Theoretical Probability **93**
22.2 Problem Solving Skill: Too Much or
Too Little Information **94**
22.4 Experimental Probability **95**

▶ **Chapter 23: Probability of Compound Events**
23.1 Problem Solving Strategy:
Make an Organized List **96**
23.2 Compound Events **97**
23.3 Independent and Dependent
Events **98**
23.4 Make Predictions **99**

▶ **Unit 8: MEASUREMENT**

▶ **Chapter 24: Units of Measure**
24.1 Algebra: Customary
Measurements **100**
24.2 Algebra: Metric
Measurements **101**
24.3 Relate Customary and
Metric **102**
24.4 Appropriate Tools and Units **103**
24.5 Problem Solving Skill: Estimate
or Find Exact Answer **104**

▶ **Chapter 25: Length and Perimeter**
25.2 Perimeter **105**
25.3 Problem Solving Strategy:
Draw a Diagram **106**
25.4 Circumference **107**

▶ **Chapter 26: Area**
   26.1 Estimate and Find Area **108**
   26.2 Algebra: Areas of Parallelograms and Trapezoids **109**
   26.4 Algebra: Areas of Circles **110**
   26.5 Algebra: Surface Areas of Prisms and Pyramids **111**

▶ **Chapter 27: Volume**
   27.1 Estimate and Find Volume **112**
   27.2 Problem Solving Strategy: Make a Model **113**
   27.3 Algebra: Volumes of Pyramids **114**
   27.5 Volumes of Cylinders **115**

▶ **Unit 9: ALGEBRA: PATTERNS AND RELATIONSHIPS**

▶ **Chapter 28: Algebra: Patterns**
   28.1 Problem Solving Strategy: Find a Pattern **116**
   28.2 Patterns in Sequences **117**
   28.3 Number Patterns and Functions **118**
   28.4 Geometric Patterns **119**

▶ **Chapter 29: Geometry and Motion**
   29.1 Transformations of Plane Figures **120**
   29.2 Tessellations **121**
   29.3 Problem Solving Strategy: Make a Model **122**
   29.4 Transformations of Solid Figures **123**
   29.5 Symmetry **124**

▶ **Chapter 30: Algebra: Graph Relationships**
   30.1 Inequalities on a Number Line **125**
   30.2 Graph on the Coordinate Plane **126**
   30.3 Graph Relations **127**
   30.4 Problem Solving Skill: Make Generalizations **128**
   30.6 Graph Transformations **129**

Name _____

LESSON 1.1

Estimate with Whole Numbers

You can use compatible numbers to estimate a quotient. **Compatible numbers** are helpful to use because they divide without a remainder, are close to the actual numbers, and are easy to compute mentally.

Oakdale Middle School is collecting recycled cans. The school has set a goal of collecting 2,788 cans. There are 38 homerooms in the school. About how many cans should each homeroom collect for the school to reach its goal?

Because you are asked "about how many," an estimate is appropriate for the answer. Use compatible numbers to estimate.

Step 1: Look at the actual numbers that make up the problem. Think about numbers that are close to the real numbers that will divide without a remainder.

2,788 ÷ 38   →   2,800 ÷ 40

Step 2: Divide.   →   2,800 ÷ 40 = 70

So, each homeroom should collect about 70 cans.

Complete to show how compatible numbers are used to estimate the quotient.

1. 2,615 ÷ 47 → 2,500 ÷ __50__ = __50__
2. 3,104 ÷ 62 → 3,000 ÷ __60__ = __50__
3. 3,591 ÷ 88 → __3,600__ ÷ 90 = __40__
4. 4,733 ÷ 74 → 4,800 ÷ __80__ = __60__
5. 7,105 ÷ 77 → __7,200__ ÷ 80 = __90__
6. 5,511 ÷ 62 → 5,400 ÷ __60__ = __90__
7. 15,843 ÷ 381 → 16,000 ÷ __400__ = __40__
8. 20,972 ÷ 287 → 21,000 ÷ __300__ = __70__
9. 29,100 ÷ 307 → __30,000__ ÷ 300 = __100__
10. 95,347 ÷ 795 → 96,000 ÷ __800__ = __120__

Use compatible numbers to estimate the quotient. *Possible estimates are given.*

11. 434 ÷ 68 __6__
12. 394 ÷ 5 __80__
13. 448 ÷ 15 __30__
14. 986 ÷ 102 __10__

15. 627 ÷ 89 __7__
16. 554 ÷ 63 __9__
17. 293 ÷ 31 __10__
18. 705 ÷ 97 __7__

19. 1,246 ÷ 43 __30__
20. 2,779 ÷ 28 __100__
21. 3,896 ÷ 38 __100__
22. 7,164 ÷ 78 __90__

Reteach **RW1**

Name _____

LESSON 1.2

Use Addition and Subtraction

Before you add or subtract, you can estimate to check for reasonableness.

Add. 599 + 462 + 630 + 348

Round to the nearest hundred.

```
   599  →     600
   462  →     500
   630  →     600
 + 348  →   + 300
              2,000
```

Now find the exact sum.

```
   599
   462
   630
 + 348
   2,039
```

Because 2,039 is close to the estimate of 2,000, the answer is reasonable.

Subtract. 1,347 − 482

Round to the nearest hundred.

```
   1,347  →    1,300
 −   482  →  −   500
                 800
```

Now find the exact difference.

```
   1,347
 −   482
     865
```

Because 865 is close to the estimate of 800, the answer is reasonable.

Complete each estimate. Then find the exact sum or difference. *Possible estimates are given.*

1.  ```
       1,325   →     1,000
       3,882   →     4,000
    + 12,356   →   + 12,000
      17,563       17,000
    ```

2.  ```
       948  →    950
     −  63  →  −  60
        885       890
    ```

3.  ```
       84,638  →    80,000
     − 36,903  →  − 40,000
       47,735       40,000
    ```

4.  ```
         524  →      500
       2,833  →    2,800
         109  →      100
     +   326  →  +   300
       3,792       3,700
    ```

Possible estimates are given.

Solve by using addition and subtraction. Estimate first.

5. The weight of the steel in the Statue of Liberty is 125 tons. The total weight of the statue is 225 tons. How much do the nonsteel materials weigh?

   100 tons; 100 tons

6. In a recent year, 19,410 U.S. students were studying in Britain, 7,872 in France, 4,715 were in Mexico, and 1,257 were studying in China. How many U.S. students were studying abroad in those four countries?

   33,000 students; 33,254 students

RW2 Reteach

Name _____

LESSON 1.3

Use Multiplication and Division

Mr. Rivera's class has collected 1,272 rocks. The students are using egg cartons to hold the rocks. If a dozen rocks fit in each carton, how many cartons are needed for the collection?

Paul did some calculations to solve this problem. This is what he did.

```
        106
    ┌───────
12 )1,272
    −12       ← (1 × 12)
    ───
     07
    − 0       ← (0 × 12)
    ───
     72
    −72       ← (6 × 12)
    ───
      0
```

- Since 12 is greater than 1, the first digit appears in the hundreds place. Paul divided the 12 hundreds.

- Paul brought down the 7 tens. Since 12 > 7, he wrote 0 in the quotient.

- Paul brought down the 2 ones. He divided the 72 ones.

The class needs 106 egg cartons to hold the collection.

Multiply or divide.

1. 399
    ×171
    68,229

2. 824
    × 32
    26,368

3. 1,440
    × 78
    112,320

4. 17)3,553   209

5. 35)7,210   206

6. 28)1,736   62

7. 38)11,666   307

8. 46)5,060   110

9. 68)34,204   503

10. 32)13,472 421

11. 43)21,973 511

12. 52)12,740 245

Reteach   RW3

Name _____

Problem Solving Strategy

Using Predict and Test

Looking for clues in a problem can help you find its answer. You can use the clues to help you guess and check different answers until you find the right one.

Valley Middle School is holding a canned food drive. Sixth-grade students have collected 150 more cans than seventh-grade students. Together, the students in both grades have collected a total of 530 cans. How many cans did the sixth graders collect? How many cans did the seventh graders collect?

Step 1: Think about what you know.
- You are asked to find the number of cans collected by each grade.
- You know the total number of cans collected and how many more cans the sixth graders collected than the seventh graders.

Step 2: Plan a strategy to solve.
- Use the *predict and test* strategy.
- Use these clues: total cans collected is 530; the difference between amounts collected by sixth and seventh graders is 150.

Step 3: Solve.
- Use a table to record your predictions and tests. Try to predict in an organized way to help you get closer to the exact answer.

PREDICT		TEST		
Sixth Graders	Seventh Graders	Clue 1: The sum is 530.	Clue 2: The difference is 150.	
330	200	330 + 200 = 530 ✓	330 − 200 = 130 ✗	← Difference is too low.
350	180	350 + 180 = 530 ✓	350 − 180 = 170 ✗	← Difference is too high.
340	190	340 + 190 = 530 ✓	340 − 190 = 150 ✓	← Both clues are satisfied.

Use the strategy *predict and test* with a table to help you solve.

1. In the problem above, what if the sixth graders had collected 120 more cans than the seventh graders? How many cans would each grade have collected?

   sixth grade: 325 cans; seventh

   grade: 205 cans

2. Tony collected 85 cans of either soup or fruit. He collected 15 more cans of soup than of fruit. How many cans of soup did he collect? How many cans of fruit did he collect?

   50 soup; 35 fruit

RW4 Reteach

Name _____

LESSON 1.5

Algebra: Use Expressions

Writing *numerical expressions* and *algebraic expressions* requires translating words into numbers and symbols. You can do this by looking for key words.

Addition	Subtraction	Multiplication	Division
sum	difference	product	quotient
increase	decrease	factors	equally divided
more than	less than	times	divided by
plus	minus	multiplied by	

- Write a numerical expression for "nine increased by seven squared."

  nine increased by seven squared
  ↓   ↓   ↓
  9   + 7^2

  A numerical expression contains only numbers.

- Write an algebraic expression for "nine increased by a number, n."

  nine increased by a number, n
  ↓   ↓   ↓
  9   + n

  An algebraic expression contains a variable.

Identify each as a *numerical expression* or *algebraic expression*. Explain your answer.

1. $3 + 6$
   numerical expression;
   no variable

2. $9 - d$
   algebraic expression;
   has a variable

3. $24 \div 12$
   numerical expression;
   no variable

Write a numerical or algebraic expression for the word expression. Remember to look for key words.

4. twenty-four less than a number, k
   $k - 24$

5. twelve increased by seventeen
   $12 + 17$

6. the product of seven and twenty
   7×20

7. thirty-eight divided by a number, m
   $38 \div m$

Reteach RW5

Name _____

LESSON 1.6

Algebra: Mental Math and Equations

You can use the number facts you know to help solve equations.
Remember, you can use fact families to find missing numbers.

- Solve the equation $8 + x = 12$ by using mental math.

  Related fact: $12 - 8 = 4$
  So, $8 + 4 = 12$.
  $8 + x = 12$
  $x = 4$ *The solution is 4.*

  Check to be sure your answer is correct.
  $8 + 4 = 12$ Replace x with 4.
  $12 = 12$ $8 + 4$ is equal to 12.

- Solve the equation $m - 9 = 8$ by using mental math.

  Related fact: $9 + 8 = 17$.
  So, $17 - 9 = 8$.
  $m - 9 = 8$
  $m = 17$ *The solution is 17.*

  Check to be sure your answer is correct.
  $17 - 9 = 8$ Replace m with 17.
  $8 = 8$ $17 - 9$ is equal to 8.

- Solve the equation $y \times 7 = 35$ by using mental math.

  Related fact: $35 \div 7 = 5$
  So, $5 \times 7 = 35$.
  $y \times 7 = 35$
  $y = 5$ *The solution is 5.*

  Check to be sure your answer is correct.
  $5 \times 7 = 35$ Replace y with 5.
  $35 = 35$ 5×7 is equal to 35.

- Solve the equation $d \div 6 = 8$ by using mental math.

  Related fact: $8 \times 6 = 48$
  So, $48 \div 6 = 8$.
  $d \div 6 = 8$
  $d = 48$ *The solution is 48.*

  Check to be sure your answer is correct.
  $48 \div 6 = 8$ Replace d with 48.
  $8 = 8$ $48 \div 6$ is equal to 8.

Solve each equation by using mental math.

1. $a - 5 = 9$
   Related fact:
   $9 + 5 =$ __14__
   The solution is __14__

2. $\dfrac{k}{5} = 5$
   Related fact:
   $5 \times 5 =$ __25__
   The solution is __25__

3. $3r = 18$
   Related fact:
   $18 \div 3 =$ __6__
   The solution is __6__

4. $10 + f = 15$
   __$f = 5$__

5. $n \div 12 = 3$
   __$n = 36$__

6. $8s = 64$
   __$s = 8$__

7. $w - 20 = 10$
   __$w = 30$__

8. $n \times 9 = 81$
   __$n = 9$__

9. $x + 9 = 16$
   __$x = 7$__

10. $\dfrac{m}{10} = 10$
    __$m = 100$__

11. $g - 30 = 6$
    __$g = 36$__

12. $16 = c \div 100$
    __$c = 1{,}600$__

RW6 Reteach

Name _____

LESSON 2.1

Use the Properties

One way to find a sum or product mentally is to use a number property.

Commutative Property
Numbers can be added in any order without changing the sum.
45 + 29 + 55 = 29 + 45 + 55
    ↑   ↑
Order has been changed.

Associative Property
Addends can be grouped differently. The sum is always the same.
(45 + 29) + 55 = 29 + (45 + 55)
              ↑   ↑
Grouping has been changed.

Numbers can be multiplied in any order without changing the product.
5 × 13 × 8 = 5 × 8 × 13
       ↑   ↑
Order has been changed.

Factors can be grouped differently. The product is always the same.
(5 × 13) × 8 = 5 × (8 × 13)
            ↑   ↑
Grouping has been changed.

Distributive Property
25 × 23 = 25 × (20 + 3) = (25 × 20) + (25 × 3)
   ↑   ↑   ↑
Product of a number   Sum of two products
and a sum

---

1. Complete to show how to find the sum. Name the reason for each step.

   19 + 45 + 21 + 5 = 19 + __21 + 45__ + 5 → __Commutative Property__

   = (__19 + 21__) + (__45 + 5__) → __Associative Property__

   = __40 + 50__ → __Use mental math.__

   = __90__

Add. Use mental math.

2. 16 + 9 + 24   3. 33 + 26 + 17 + 44   4. 21 + 14 + 29 + 36

   __49__   __120__   __100__

Complete to show how to use the Distributive Property to find each product.

5. 8 × 14 = 8 × (__10__ + 4)

   = (8 × __10__) + (8 × __4__)

   = __80__ + __32__

   = __112__

6. 9 × 34 = 9 × (30 + __4__)

   = (9 × __30__) + (9 × __4__)

   = __270__ + __36__

   = __306__

Reteach RW7

Name _____

LESSON 2.2

Exponents

Powers of numbers can be written in exponent form. An *exponent* shows how many times a number called the *base* is used as a factor.

$10^5 = \underbrace{10 \times 10 \times 10 \times 10 \times 10}_{\text{factors}} = 100{,}000$

base — 10 used as a factor 5 times

$2^6 = \underbrace{2 \times 2 \times 2 \times 2 \times 2 \times 2}_{\text{factors}} = 64$

base — 2 used as a factor 6 times

Joan is asked to express 64 using an exponent and the base 4.

Step 1 The base is 4. So, Joan must find equal factors.

$8 \times 8 = 64$
$4 \times 2 \times 2 \times 4 = 64$
$4 \times 4 \times 4 = 64$

Step 2 Joan writes the base. Then she counts how many times it is used as a factor.

$4^3 \leftarrow$ used as a factor 3 times
↑
base

So, $64 = 4^3$.

Write in exponent form.

1. $4 \times 4 \times 4$ ___4^3___
2. $3 \times 3 \times 3 \times 3 \times 3$ ___3^5___
3. $10 \times 10 \times 10 \times 10 \times 10 \times 10$ ___10^6___
4. $30 \times 30 \times 30 \times 30$ ___30^4___

Express with an exponent and the base given.

5. 36, base 6

___6^2___

6. 32, base 2

___2^5___

7. 81, base 3

___3^4___

8. 625, base 5

___5^4___

9. 343, base 7

___7^3___

10. 128, base 2

___2^7___

11. 512, base 8

___8^3___

12. 1,000,000, base 10

___10^6___

13. 121, base 11

___11^2___

RW8 Reteach

Name _____

LESSON 2.4

Order of Operations

When an expression involves more than one operation, you use the order of operations to evaluate it.

1. Operate inside parentheses.
2. Clear exponents.
3. Multiply and divide from left to right.
4. Add and subtract from left to right.

What is the value of $2 \times 3 + (6 - 2) \div 2$?

Step 1 Simplify inside parentheses.
Step 2 There are no exponents.
Step 3 Multiply and divide in order from left to right.
Step 4 Add and subtract in order from left to right.

$$2 \times 3 + (6-2) \div 2$$
$$2 \times 3 + 4 \div 2$$
$$6 + 2$$
$$8$$

Evaluate the expression.

1. $5 + 4 + 7 \times 3$

   $5 + 4 + \underline{\ 21\ }$

   $\underline{\ 9\ } + \underline{\ 21\ }$

   $\underline{\ 30\ }$

2. $28 \div 4 - (4 + 3)$

   $28 \div 4 - \underline{\ 7\ }$

   $\underline{\ 7\ } - \underline{\ 7\ }$

   $\underline{\ 0\ }$

3. $2^2 + 16 - 8$

   $\underline{\ 4\ } + 16 - 8$

   $\underline{\ 20\ } - 8$

   $\underline{\ 12\ }$

4. $(6 + 4)^2 \times 8$

   $(\underline{\ 10\ })^2 \times 8$

   $\underline{\ 100\ } \times 8$

   $\underline{\ 800\ }$

Find the value of each expression. From the box at the right, choose the letter that corresponds to the value.

5. $(5 \times 6) \div (4 + 2)$ __B__

6. $10 + 15 - 9 \times 2$ __C__

7. $3^2 + 1 - (18 - 9)$ __A__

8. $(4 + 8) \times 3 \div 6 + 10$ __E__

9. $2 + 5^2 - (4 + 3)$ __G__

10. $4 \div 2 \times 6 - 3 + 3^2$ __F__

A	1
B	5
C	7
D	10
E	16
F	18
G	20
H	32

Reteach **RW9**

Name _____

LESSON 2.5

Problem Solving Skill

Sequence and Prioritize Information

Brian wants to bake some cookies for his class party. He knows he will need to mix the ingredients, preheat the oven, assemble the ingredients, cool the cookies, and bake the cookies, but he is not sure how to begin. The order in which Brian does the steps is important. You can help Brian decide what to do.

Here are the steps he needs to do:

Mix the ingredients.
Preheat the oven.
Assemble the ingredients.
Cool the cookies.
Bake the cookies.

In order to help Brian, write the steps on 5 index cards, with one step on each card.

Now put the steps in order, by changing the order of the index cards until the sequence of steps makes sense.

1. Preheat the oven; 2. Assemble the ingredients; 3. Mix
the ingredients; 4. Bake the cookies; 5. Cool the cookies.

Write the steps in order.

1. To make mashed potatoes:
   - Drain the potatoes.
   - Mash the potatoes.
   - Peel the potatoes.
   - Add butter and milk, and stir.
   - Cook the potatoes.

   1. _Peel the potatoes._
   2. _Cook the potatoes._
   3. _Drain the potatoes._
   4. _Mash the potatoes._
   5. _Add butter and milk, and stir._

2. To fly to another city:
   - Get on the plane.
   - Buy a ticket.
   - Check in at the gate.
   - Drive to the airport.

   1. _Buy a ticket._
   2. _Drive to the airport._
   3. _Check in at the gate._
   4. _Get on the plane._

3. To find the value of (3 × 5) + 2 − 8:
   - Subtract 8.
   - Multiply 3 by 5.
   - Add 2.
   - What is your answer? ____9____

   1. _Multiply 3 by 5._
   2. _Add 2._
   3. _Subtract 8._

RW10   Reteach

Name _____

LESSON 3.1

Represent, Compare, and Order Decimals

The numbers you use every day are part of the decimal system. To find the value of a number, look at the digits and the position of each digit. A place-value chart can help you.

Place Value									
Millions	Hundred Thousands	Ten Thousands	Thousands	Hundreds	Tens	Ones	Tenths	Hundredths	Thousandths	Ten-Thousandths

↑ decimal point

Knowing place values is useful when comparing decimal numbers. Mindy is asked to list 18.3, 17.8, and 24.1 in order from greatest to least.

Step 1 Mindy compares the first two numbers.
- She starts at the left. Both numbers have the digit 1 in the tens place.   1̲8.3 ↔ 1̲7.8
- So, Mindy looks at the digits in the ones place. The first number has the digit 8 in the ones place, while the second number has the digit 7.   18̲.3 ↔ 17̲.8
- Since 8 > 7, 18.3 > 17.8.

Step 2 Mindy compares the third number to the greatest number so far, the first number.   2̲4.1 ↔ 1̲8.3
- The third number has the digit 2 in the tens place, while the first number has the digit 1 in the tens place.
- Since 2 > 1, 24.1 > 18.3.

Using what she has discovered, Mindy makes the list: 24.1; 18.3; 17.8.

Give the position and the value of each underlined digit.

1. 30.1̲94
   a. position _____
   b. value _____

2. 4,082,113.723̲
   a. position _____
   b. value _____

Compare the numbers. Write <, >, or = in the ●.

3. 9.03 ● 0.93   4. 0.210 ● 0.012   5. 8.241 ● 8.24

Write the numbers in order from least to greatest.

6. 42.05; 45.02; 40.52   7. 19.7; 19.007; 19.07   8. 0.59; 0.95; 0.6

   _____   _____   _____

Reteach **RW11**

Name _____

LESSON 3.2

Problem Solving Strategy: Make a Table

Putting data in numerical order in a table can often help you determine the greatest or the least piece of data.

The areas of some sports fields are given below.

Basketball . . . . .427 square yards   Ice hockey . . . . .2,222 square yards
Football . . . . .6,400 square yards   Tennis (doubles) . . . . . .312 square yards

The area of an Olympic swimming pool is 5,135 square yards less than the greatest area above and 953 square yards greater than the least area. What is the area of an Olympic swimming pool?

Step 1 Think about what you know and what you are asked to find.
- You know the sizes of different sports fields.
- You need to find the area of the pool.

Step 2 Plan a strategy to solve.
- Use the strategy *make a table* to order the data.
- Use the table entries to find the area of the pool.

Step 3 Carry out the strategy.

The greatest area is 6,400 square yards and the least area is 312 square yards. Use the greatest area to find the area of an Olympic swimming pool.

Tennis (doubles)	312 square yards
Basketball	427 square yards
Ice hockey	2,222 square yards
Football	6,400 square yards

6,400 − 5,135 = 1,265

So, the area of an Olympic pool is 1,265 square yards.

Now find the difference between the area of an Olympic swimming pool and the least area to check your answer.

1,265 − 312 = 953   The answer checks.

Solve the problem by making a table.

1. The areas of some other sports fields are boxing, 44 yd^2; fencing, 33 yd^2; judo, 306 yd^2; karate, 75 yd^2; and kendo, 132 yd^2. The area used for wrestling is 10 yd^2 greater than the least area above. How does the area for wrestling compare to the area for karate?

    The area for karate is 32 yd^2 greater.

    _____

2. Baseball fields are not a standard size. However, the bases on the field are the corners of a square whose area is 5,500 yd^2 less than the area of a football field and 867 yd^2 greater than the area for a fencing match. What is the area of this square?

    900 yd^2

    _____

RW12   Reteach

Name _____

LESSON 3.3

Estimate with Decimals

When you estimate with decimals, you want to get an idea of the size of the result. Working with whole numbers can help you estimate quickly.

Estimate 19.7 + 40.13 + 100.4.

Step 1 Round each number to the nearest whole number.

19.7 → 20
40.13 → 40
100.4 → 100

Step 2 Add the rounded numbers.

20 + 40 + 100 = 160

So, a good estimate for the sum is 160.

Estimate $78.31 − $49.47.

Step 1 Round to the nearest ten.

$78.31 → $80
$49.47 → $50

Step 2 Subtract the rounded numbers.

$80 − $50 = $30

So, a good estimate for the difference is $30.

Estimate 62.88 × 28.97

Step 1 Round to the nearest ten.

62.88 → 60
28.97 → 30

Step 2 Multiply the rounded numbers.

60 × 30 = 1,800

So, a good estimate for the product is 1,800.

Estimate 54.67 ÷ 8.56.

Step 1 Find compatible numbers close to those given.

54.67 → 54
8.56 → 9

Step 2 Divide the compatible numbers.

54 ÷ 9 = 6

So, a good estimate for the quotient is 6.

Estimate. *Possible estimates are given.*

1. 19.82 + 51.5 __70__
2. 149.2 ÷ 23.8 __6__
3. 784.49 − 610.88 __200__
4. 39.66 × 6.75 __280__
5. 1003.2 − 796.1 __200__
6. 7.86 + 10.03 __18__
7. 98.15 × 8.23 __800__
8. 82.88 ÷ 9.31 __9__
9. 38.8 × 9.12 __360__
10. 161.10 ÷ 7.84 __20__
11. 108.46 + 392.54 __500__
12. 80.55 − 67.86 __10__
13. 57.93 × 21.5 __1,200__
14. 119.4 ÷ 42.3 __3__
15. 53.3 + 39.2 __90__
16. 48.28 ÷ 6.82 __8__
17. 28.7 × 61.75 __1,800__
18. 982.3 − 498.7 __500__
19. 411.9 + 298.34 + 128.6 __800__
20. $49.28 + $32.61 + $18.95 __$100__

Reteach **RW13**

Name _____

LESSON 3.4

Decimals and Percents

If a figure is divided into 100 equal parts, each part represents $\frac{1}{100}$, 0.01, or 1% of the whole.

The shaded squares in the 10 × 10 grid at the right represent $\frac{54}{100}$, 0.54, or 54% of the whole square.

Percent can also represent value.

One hundredth of the value of a dollar is 1% of its value. So, one cent is 1% of one dollar.

$0.43 represents 0.43, or 43% of one dollar.

$0.70 represents 0.70, or 70% of one dollar.

Write the decimal and percent for the amount of money as part of one dollar.

1.

   _0.18, 18%_

2.

   _0.82, 82%_

3.

   _0.60 or 0.6, 60%_

Write the corresponding decimal or percent.

4. 75%   5. 6%   6. 0.15   7. 0.40

   _0.75_   _0.06_   _15%_   _40%_

8. 71%   9. 0.08   10. 98%   11. 0.12

   _0.71_   _8%_   _0.98_   _12%_

RW14 Reteach

Name _____

LESSON 4.1

Add and Subtract Decimals

Ian is buying school supplies.
Find the total cost.

Binder	$3.49	Paper	$2.79
Pen	$0.88	Highlighter	$0.98

Step 1: List the items in a column.
Remember to align the decimal points.

$3.49
0.88
2.79
+ 0.98

Step 2: Write the decimal point for the answer.
Place it directly under the other points.

Step 3: Add as you would whole numbers.
Remember to regroup if needed.

$3.49
0.88
2.79
+ 0.98
$8.14

The total cost of Ian's school supplies is $8.14.

Now suppose Ian pays for his supplies with a $20 bill.
Find the amount of change he should receive.

Step 1: Write the numbers in a column.
Remember to align the decimal points.

$20.00
− 8.14

Step 2: Write the decimal point for the answer.
Then subtract as you would with whole
numbers, regrouping as necessary.

$20.00
− 8.14
$11.86

Ian should receive $11.86 in change.

Add or subtract.

1. $75.50 − $47.86

   $27.64

2. 347.9 − 69.38

   278.52

3. 81.42 − 57.932

   23.488

4. 2.89 + 1.65 + 3.86

   8.4

5. 4.62 + 7.89 + 9.17

   21.68

6. 2.891 + 3.006 + 2.861

   8.758

7. 18.21 + 6.85 + 2.77

   27.83

8. 4.0689 − 1.0791

   2.9898

9. 2.478 + 6.811 + 7.222

   16.511

10. 6.42 + 5.1 + 0.28

    11.8

11. 3.016 − 1.2173

    1.7987

12. 38.2 + 5 + 6.83

    50.03

Reteach **RW15**

Name _____

LESSON 4.2

Multiply Decimals

Marty worked 26.5 hours this week. He earns $6.40 per hour. How much money did Marty earn this week?

To solve, you need to find the product 26.5 × 6.40.

Step 1: Multiply as you would with whole numbers.

$6.40
× 26.5

169600

Step 2: Count the number of decimal places in the factors.

$6.40 → 2 decimal places
× 26.5 → 1 decimal place

169600 → 3 decimal places

Step 3: Starting at the right side of the answer, count over that number of places. This is where the decimal point is placed.

$6.40
× 26.5

$169.600

Marty earned $169.60 this week.

Complete to solve.

1. Sean worked 34 hours this week. He earns $9.25 an hour. How much money did Sean earn this week?
   a. What multiplication problem will you use? ____34 × $9.25____
   b. How many decimal places are in the factors? ____2____
   c. How much did Sean earn? ____$314.50____

2. Denise earns $4.65 an hour. Last week she worked 17 hours. How much did she earn last week?
   a. What multiplication problem will you use? ____17 × $4.65____
   b. How many decimal places are in the factors? ____2____
   c. How much did Denise earn? ____$79.05____

Multiply.

3. 23.1 × 5.7
   ____131.67____

4. 28.44 × 7
   ____199.08____

5. 25.4 × 24.55
   ____623.57____

6. 16.6 × 0.24
   ____3.984____

7. 101.01 × 8.8
   ____888.888____

8. 2.32 × 11.48
   ____26.6336____

RW16 Reteach

Name _____

LESSON 4.4

Divide with Decimals

Mary has saved $0.35 each day. She now has a total of $9.10. How long has Mary been saving?

To solve, you need to divide 9.10 by 0.35.

Step 1: Make the divisor a whole number by multiplying the divisor by a multiple of 10. Multiply the dividend by the same multiple of 10.

$0.35 \overline{)9.10}$ → $35 \overline{)910}$
× 100 × 100

Step 2: Place the decimal point in the quotient directly above the decimal point in the dividend. Divide as you would with whole numbers.

$$\begin{array}{r} 26. \\ 35\overline{)910.} \\ -70 \\ \hline 210 \\ -210 \\ \hline 0 \end{array}$$

Since the remainder is 0, the answer is a whole number. You do not need to show the decimal point.

Mary has been saving for 26 days.

Complete to solve.

1. Dusty bought 7 movie tickets for his friends. The tickets cost a total of $34.65. How much does each friend owe him for the cost of one ticket?

   a. What division problem will you use to solve? _____$34.65 ÷ 7_____

   b. Do you need to multiply to make the divisor a whole number? _____no_____

   c. How do you decide where to place the decimal point in the quotient? __Place it directly above the decimal point in the dividend.__

   d. How much does each friend owe? _____$4.95_____

Place the decimal point in the quotient.

2. 131.52 ÷ 6.4 = 2055   3. 50.085 ÷ 10.6 = 4725   4. 1936.95 ÷ 2.22 = 8725

   ____20.55____   ____4.725____   ____872.5____

Find the quotient.

5. 16.88 ÷ 5   6. 81.9 ÷ 18   7. 332.8 ÷ 40

   ____3.376____   ____4.55____   ____8.32____

8. 118 ÷ 12.5   9. 203.205 ÷ 5.7   10. 421.155 ÷ 14.7

   ____9.44____   ____35.65____   ____28.65____

Reteach RW17

Name _____

> LESSON 4.5

Problem Solving Skill: Interpret the Remainder

When you solve a problem using division, there is often a remainder. When there is a remainder, you must decide what it means for the problem. You need to determine whether the solution to the problem is
- the quotient without the remainder,
- the next whole number greater than the quotient,
- the next whole number less than the quotient, or
- the remainder.

A real-estate agent ordered 190 doughnuts for an open house. The Donut Shoppe packed the order in boxes of 12 doughnuts. How many boxes were needed?

Step 1: Think about what you know and what you are asked to find.
- You know the number of doughnuts ordered and the number that were put in each box.
- You are asked to find the number of boxes that were needed.

Step 2: Decide on a plan to solve.
- Since the same number of doughnuts were put into each box, use division to solve the problem.
- Examine the quotient and remainder and decide how they relate to the problem.

Step 3: Carry out the plan.

$$12 \overline{)190} = 15 \text{ r } 10$$

The quotient 15 means that 15 boxes were filled. The remainder 10 means that there were 10 doughnuts left after the last full box was packed.

So, the Donut Shoppe needed 15 + 1, or 16 boxes in all, to pack the agent's 190 doughnuts.

Interpret the remainder to solve.

1. A school received a shipment of 166 new social studies textbooks. The textbooks had been packed into cartons with 12 books in each full carton. How many cartons were delivered to the school?
   14 cartons (13 full cartons, 1 with 10 books)

2. In the school science laboratory, students generally work in groups of 4. Extra students are divided among the groups to make some groups of 5. If there are 27 students in a class, how many groups will have 5 students?
   3 groups (out of 6)

RW18 Reteach

Name _____

LESSON 4.6

Algebra: Decimal Expressions and Equations

You speak and write using words, for example

> the number of days in a week times the number of weeks

Some words can be expressed as numbers or numerical expressions.

> days × weeks
> 7 × 5

An algebraic expression uses a letter to represent a number. For example, if the number of weeks is written as the letter w, the algebraic expression is $7 \times w$.

It can also be written as $7w$.

To evaluate the expression $7w$ to find the number of days in 8 weeks, write $7 \times 8 = 56$.

Match the algebraic expression with the words.

1. number of inches in a foot times the number of feet ___C___
2. number of items in a dozen plus some more items ___A___
3. sum of money divided among several people ___D___
4. total amount of money less the amount spent ___B___

A	$12 + s$
B	$\$25 - d$
C	$12n$
D	$\$25 \div p$

Solve using mental math.

5. How much is $6y$ if $y = 3$? __18__
6. How much is $8 + r$ if $r = 9$? __17__
7. How much is $24 \div t$ if $t = 3$? __8__
8. How much is $35 - a$ if $a = 15$? __20__
9. How much is $4.5 - q$ if $q = 1.9$? __2.6__
10. How much is $5b$ if $b = 1.1$? __5.5__
11. How much is $3.6 \div z$ if $z = 3$? __1.2__
12. How much is $2.4 + c$ if $c = 4.2$? __6.6__

Evaluate each expression.

13. $5m$ for $m = 3.3$
   __16.5__
14. $w \div 2$ for $w = 8.4$
   __4.2__
15. $y + 12.4$ for $y = 5.2$
   __17.6__
16. $7.4 - b$ for $b = 1.3$
   __6.1__
17. $6x$ for $x = 2.5$
   __15__
18. $m - 4.6$ for $m = 8.9$
   __4.3__

Reteach RW19

Name _____

LESSON 5.1

Samples

A sample is a part of a population. Samples are used when it would be too time-consuming or too expensive to survey an entire population.

A population may be people, such as all the people in a city, or it may be objects, such as all the books in a bookstore.

A sample is chosen using one of several methods. Three such methods are:

- **Random sample:** Each person or object in a population has the same chance of being chosen.
- **Convenience sample:** Those people or objects that are most readily available are chosen.
- **Systematic sample:** A person or object is chosen randomly and then a pattern is used, such as every fifth person, to choose others.

In general, it is best to use a random sample, since it is most representative of an entire population. It is also important to use the largest sample possible, since the larger the sample, the closer the results will represent the entire population.

---

1. Carole is surveying members of her community asking them to name their favorite restaurant. Determine the type of sample. Write *convenience, random,* or *systematic.*

   a. She stands outside a local mall and asks people leaving the mall to name their favorite restaurant.

   _____convenience_____

   b. She opens the telephone book, randomly chooses a name, and then chooses every tenth name that begins with the same letter.

   _____systematic_____

   c. She knows that every family in her community is listed in the telephone book. She calls 5 people randomly chosen from each page of the book.

   _____random_____

2. Evan is on a committee to determine if the players in the Tri-Town Soccer League want to extend the season. Tell whether he should survey *all the players* or use *a sample* in each situation.

   a. There are 40 players in the league.   _____all the players_____

   b. The players are difficult to reach since they live in many areas.   _____a sample_____

   c. There are 650 players in the league.   _____a sample_____

RW20   Reteach

Name _____

LESSON 5.2

Bias in Surveys

A sample is **biased** if individuals or groups from the population are not represented in the sample.

Linda wanted to survey the 125 sixth-grade students at her school to find out the number of hours they spent reading each week. She made a list of four sampling methods she could use.

- randomly survey 5 students
- randomly survey 15 sixth-grade boys
- randomly survey 13 students at her school
- randomly survey 15 sixth-grade students

She decided that the first method would not include a large-enough sample. The second method would not include sixth-grade girls. The third method was biased because it would not be exclusively for sixth graders. Linda decided to use the last method. It would contain a large-enough sample, and every member of the sixth grade would have an equal chance of being selected.

---

1. Marian wants to survey the 58 sixth-grade teachers at Oak Park School to find out the average number of hours they spend grading papers. She makes a list of four sampling methods she could use. Circle the sampling method that is not biased. Then explain how the other three methods are biased.

   a. randomly survey all teachers who have taught for more than 10 years
   b. randomly survey 10 male teachers at the school
   c. randomly survey 8 sixth-grade teachers ⟵ circled
   d. randomly survey 8 math teachers at the school

   Method a excludes teachers who have taught less than 10 years. Method b excludes
   female teachers. Method d excludes teachers who teach subjects other than math.

2. Dan wants to survey the 230 members of a golf club to find out the average number of hours they play golf each week. He makes a list of four sampling methods he could use. Circle the sampling method that is not biased. Then explain how the other three methods are biased.

   a. randomly survey all club members who have won tournaments
   b. randomly survey 30 club members ⟵ circled
   c. randomly survey all members who have ever shot a hole-in-one
   d. randomly survey 25 members under the age of 30

   Method a excludes members who have not won tournaments. Method c excludes
   members who never shot a hole-in-one. Method d excludes members over the age of 30.

Name _____

LESSON 5.3

Problem Solving Strategy: Make a Table

Sometimes, when you have a lot of data, you do not need to know about each value. It may be enough to know how many values fit into a particular category. When this is the case, a tally table can help you solve problems.

Mr. Franks corrected the test papers of the 29 students in his 3rd period class. The scores (out of 100) are given below. How many more students had a score below 90 than had a score of 90 or above?

73, 83, 85, 92, 93, 85, 89, 91, 99, 80, 80, 84, 76, 78, 80, 93, 90, 88, 98, 82, 100, 78, 67, 88, 98, 94, 90, 76, 73

Step 1: Think about what you know and what you are asked to find.
- You know all the grades that the students in the 3rd period class received. You do not know who received each grade, but that information is not needed in this situation.
- You are asked to find the difference between the number of students who scored less than 90 and the number who scored 90 or above.

Step 2: Plan a strategy to solve.
- Use the strategy *make a table*
- Organize the data into a tally table. Make only as many rows as you need to solve the problem. For this problem, you need 2 rows: *less than 90* and *90 or above*.

Step 3: Solve.
- Carry out the strategy.

Test Scores	
Less than 90	|||| |||| |||| |||
90 or above	|||| ||||	

Count the tallies and find the difference. 18 − 11 = 7

So, 7 more students had a score below 90 than had a score of 90 or above.

Use the data below and the strategy *make a table* to help you solve 1–2.

The students in a sixth-grade class were surveyed about the number of minutes they spend on the Internet on a typical day. The results of the survey are given below.

15, 0, 20, 15, 30, 0, 0, 0, 15, 20, 60, 45, 20, 10, 0, 30, 0, 0, 90, 45, 0, 10, 0, 20

1. How many more students said they spend some time on the Internet each day than said they do not spend any time on the Internet?

   _____ 6 more students _____

2. Of those students who use the Internet, how many more of them spend less than 30 minutes online than spend at least 30 minutes online?

   _____ 3 more students _____

RW22 Reteach

Name _____

LESSON 5.4

Frequency Tables and Line Plots

A local fitness center wants to survey 50 adults to find out if they exercise daily. The age of each person who said *yes* was recorded. Use the data below to make a line plot.

Ages of Adults Who Said *Yes*						
28	33	45	25	50	42	33
25	48	31	37	28	25	50
42	29	45	50	38	31	29
38	52	40	33	25	37	52

Step 1: Draw a horizontal line.
Step 2: On the line, write numerical values for the ages, using vertical tick marks.
Step 3: Plot the data by drawing an x on the line plot for each value, or person's age, in the table.

```
X
X                                                                    X
X     X X     X    X         X X        X        X             X     X
X     X X     X    X         X X    X   X        X        X    X     X
+--+--+--+--+--+--+--+--+--+--+--+--+--+--+--+--+--+--+--+--+--+--+--+--+--+--+--+--+
25 26 27 28 29 30 31 32 33 34 35 36 37 38 39 40 41 42 43 44 45 46 47 48 49 50 51 52
```

For Exercises 1–2, use the data in the tables to make a line plot.

1. A local bookstore wants to survey 50 students to find out if they buy at least one book a month. The surveyor recorded the age of each student who said *yes*.

Ages of Students Who Said *Yes*						
13	12	16	15	11	14	15
11	16	14	14	13	15	11
14	15	11	12	16	11	13

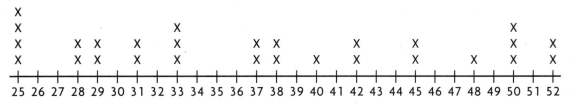

2. The 20 students in science class take a test. Here are their results.

Science Test Scores				
76	82	85	95	98
92	78	76	90	85
88	95	74	78	76
74	85	92	82	88

```
                  X                    X
         X        X        X        X        X     X        X
X    X   X    X   X    X   X   X    X   X    X     X   X    X   X
+--+--+--+--+--+--+--+--+--+--+--+--+--+--+--+--+--+--+--+--+--+--+--+--+--+
74 75 76 77 78 79 80 81 82 83 84 85 86 87 88 89 90 91 92 93 94 95 96 97 98
```

Reteach   RW23

LESSON 5.5

Name _____

Measures of Central Tendency

The test scores of a math class are shown below. Find the mean, median, and mode for the data.

Test Scores									
78	81	93	91	100	100	81	78	98	100

Mean: Find the sum of the scores and divide by the number of scores.

$$78 + 81 + 93 + 91 + 100 + 100 + 81 + 78 + 98 + 100 = 900$$
$$900 \div 10 = 90$$

Median: Arrange the scores in order from least to greatest. Find the number that is in the middle. Since there is an even number of scores, find the mean of two middle numbers.

78 78 81 81 <u>91 93</u> 98 100 100 100

92 ← The mean of 91 and 93 is 92.

Mode: Find the number or numbers that appear most often.

100

So, the mean test score is 90. The median score is 92. The mode score is 100.

Find the mean, median, and mode for each set of data.

1.
Test Scores									
85	77	91	97	95	95	79	83	95	93

mean: ___89___

median: ___92___

mode: ___95___

2.
Golf Scores						
88	76	92	91	69	76	

mean: ___82___

median: ___82___

mode: ___76___

3.
Number of School Days								
Japan	England	Israel	Germany	Netherlands	USA	Thailand	Sweden	Canada
243	200	216	210	200	180	200	180	180

mean: ___201___   median: ___200___   mode: ___180 and 200___

RW24 Reteach

Name _____

LESSON 5.6

Outliers and Additional Data

The measures of central tendency help you describe a set of data using only one number. Three measures of central tendency are the *mean*, the *median*, and the *mode*. When additional data values are added to a set, one or more of these measures may change.

Jenna has the five items shown below in her shopping cart.

She can describe the average amount in the containers using each of the measures of central tendency.

Mean: $(8 + 6 + 16 + 12 + 12) \div 5 = 54 \div 5 = 10.8$ ounces
Median: 6   8   **12**   12   16   →   12 ounces
Mode: 12 ounces

When Jenna adds the container of juice at the right to her cart, one or more of the measures of central tendency may change

Mean: $(8 + 6 + 16 + 12 + 12 + 64) \div 6 = 118 \div 6 \approx 19.7$ ounces
Median: 6   8   **12   12**   16   64   → $(12 + 12) \div 2 = 12$ ounces
Mode: 12 ounces

When the large container was added to the others in the cart, only the mean changed. The median and mode remained the same as before.

When new data values are added to a set, often the mean changes the most because the mean is found by adding all the data values.

The median and mode usually are not affected as greatly by the addition of an *outlier*, a data value very different from the other data values in a set.

The table below shows how much money Garrett spent buying house plants. Use the data for 1–2.

Date	March 1	March 7	March 15	March 24
Amount Spent	$10.00	$20.00	$15.00	$15.00

1. Find the mean, median, and mode of the amounts.

   _____mean: $15; median: $15; mode: $15_____

2. On April 3, Garrett bought a large plant that cost $60. What are the new mean, median, and mode for the costs of the five plants?

   _____mean: $24; median: $15; mode: $15_____

Reteach   RW25

Name _____

LESSON 5.7

Data and Conclusions

You must consider many things when you are deciding whether a conclusion based on a set of data is valid. Here are four considerations.

- You must know the population that you are interested in. A population may be people, such as the students in your school or class, or it may be objects, such as the cars produced by a factory in one week.
- You must know whether that population was surveyed. For example, if you are interested only in adults between the ages of 25 and 40, how do you know that only people of this age were surveyed?
- You must know whether each survey question was unbiased. Were people led to believe that you wanted one answer rather than another?
- Were the people or objects selected randomly so that every member of the population had an equal chance of being chosen?

Diana's school has 986 students attending it. She asked 80 randomly selected students in the school the question "What kind of pet or pets do you have?" The results of her survey were as follows:

    26% said they had a cat.

    23% said they had a dog.

    12% said they had a pet other than a cat or dog.

She concluded that percentages close to these would apply to all the students in her school. Her conclusion was valid because:

- She wanted to find out data about the entire population of her school.
- She surveyed the population she was interested in.
- She asked the question in a fair and unbiased way.
- The sample was large enough to be representative of the students in her school and the students sampled were selected randomly.

Write *yes* or *no* to tell whether Diana's conclusions would have been valid if she had used each of these surveying methods. Explain your answer.

1. Diana asked some students who attend her school, as well as some students who attend another middle school.

   No, the students at the other school would not be

   representative of the students at her school.

2. Diana asked the first 80 students on the daily attendance sheet, which has students' names listed alphabetically.

   Yes. Students' last names are not related to their having pets,

   so her conclusions would still be valid.

RW26 Reteach

Name _____

LESSON 6.1

Make and Analyze Graphs

At Fox Run School, the sixth-grade and seventh-grade classes are competing to see which class can collect more aluminum cans. The bar graphs show the results for the first three weeks.

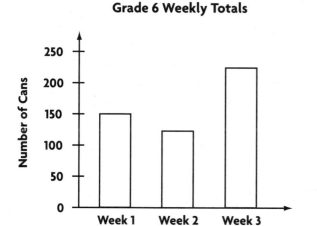

To compare the two classes' results more easily, you can show all the data in one graph. The following graph is a double-bar graph.

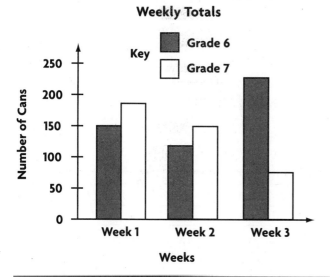

The key tells which bar represents the sixth grade and which represents the seventh grade. A bar graph that shows two or more sets of data is called a multiple-bar graph.

For Exercises 1–3, use the double-bar graph above.

1. How do you know which bar refers to which grade? _____ Read the key _____

2. Which grade collected more cans in week 1? in week 2? in week 3?

    _____ Grade 7; Grade 7; Grade 6 _____

3. Which grade collected more cans altogether? _____ Grade 6 _____

Reteach **RW27**

Name _____

LESSON 6.2

Find Unknown Values

Line graphs often show a pattern. You can use a line graph to estimate unknown values by extending the graph.

Shauna swims laps in a pool to train for a race. She swims at a rate of 25 m per min. The table below shows how far she swims.

Time (min)	1	2	3	4	5
Distance (m)	25	50	75	100	125

Make a line graph from the data in the table. Use the graph to estimate how many minutes it will take Shauna to swim 150 m.

Step 1 Use the data to draw the graph.

Step 2 Extend the graph until it crosses the 150 m line. Look directly down to the horizontal axis and estimate the value along that axis. If Shauna continues to swim at this rate, it will take her about 6 min to swim 150 m.

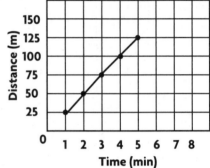
Shauna's Race Training

You can also use logical reasoning and arithmetic. Since 150 m = 100 m + 50 m, add the number of minutes she takes to swim those two distances: 4 min + 2 min = 6 min.

The formula $d = rt$, where d is the distance, r is the rate, and t is the time, can also be used to find the time Shauna needs to swim 150 m.

$d = rt$ Write the formula.
$150 = 25t$ Substitute the values you know into the formula.
$6 = t$ Solve to find the value of t.

So, it will take Shauna 6 min to swim 150 m.

A bicycle racer has averaged 20 mi per hr during the first 4 hr of a race.

Time (hr)	1	2	3	4
Distance (mi)	20	40	60	80

1. Make a line graph. Use the graph to estimate how long it will take the racer to ride 100 mi. __Check students' graphs; about 5 hr.__

2. Use logical reasoning and arithmetic to find how long it will take the racer to ride 100 mi. __5 hr__

3. Use the formula $d = rt$ to find how long it will take the racer to ride 140 mi.

__7 hr__

Name _____

LESSON 6.3

Stem-and-Leaf Plots and Histograms

A histogram is a type of bar graph. Histograms are different from bar graphs in two ways.
- The bars are side by side, not spaced apart.
- Each bar refers to an interval, not a single item.

All of the students in Mr. Higgins' physical education class ran the 100-yard dash. This histogram shows the students' times to the nearest second.

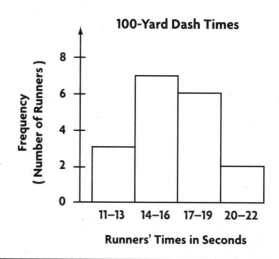

Complete.

1. The number of runners with times of 17–19 sec was ____6____.

2. ____3____ students had times of 11–13 sec.

There were 18 students who recorded how many baskets they made in 30 sec. Their results are in the following tables.

Student	1	2	3	4	5	6	7	8	9
Baskets made	5	8	17	13	15	5	19	7	7

Student	10	11	12	13	14	15	16	17	18
Baskets made	8	9	6	14	7	16	12	12	10

3. Use the data above to complete the histogram.

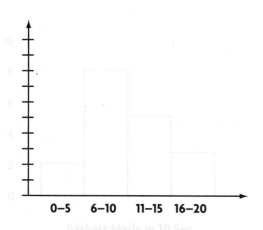

4. Complete the stem-and-leaf plot of the Baskets Made data.

Stem	Leaves
0	5 5 6 7 7 7 8 8 9
1	0 2 2 3 4 5 6 7 9

Key: 0 | 5 = 5

Reteach RW29

Name _____

LESSON 6.5

Box-and-Whisker Graphs

The owner of a gift shop made a box-and-whisker graph to represent the number of customers who came into the shop each day.

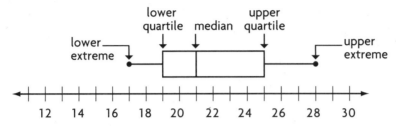

The least number in a box-and-whisker graph is called the **lower extreme.** It is the black dot at the end of the left whisker. In this graph, the lower extreme represents the least number of customers to enter the shop in a single day. The lower extreme is 17.

The greatest number in a box-and-whisker graph is called the **upper extreme.** It is the black dot at the end of the right whisker. In this graph, the upper extreme represents the greatest number of customers to enter the shop in one day. The upper extreme is 28.

The range is the difference between the upper and lower extremes. The range of this set of data is 11.

The median of a set of data is the middle number when all numbers are arranged in numerical order. The median of this set is 21.

Another store owner also made a box-and-whisker graph of the customers who entered her store daily.

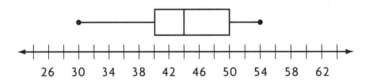

1. What is the lower extreme? ____30____

2. What is the upper extreme? ____54____

3. What is the lower quartile? ____40____

4. What is the upper quartile? ____50____

5. What is the range of this set of data? ____24____

6. What is the median for this set of data? ____44____

RW30 Reteach

Name _____

LESSON 6.6

Analyze Graphs

Graphs are sometimes drawn in order to mislead the reader. In order for a graph to provide information honestly, it must meet several requirements. One requirement is that the scale must be accurate.

Stereo City placed an advertisement for a stereo system in a newspaper. The advertisement included the graph at the right, which compares its price for the system to the price of the system at Speaker Town.

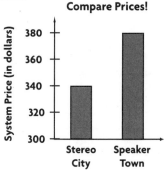

In the graph, the bar for Speaker Town appears to be twice the height of the bar for Stereo City. Some readers might think this means Speaker Town's price is twice that of Stereo City's. The scale, however, shows that Speaker Town's price for the stereo system is not twice the price at Stereo City. The actual difference between the two prices is only $380 - $340, or $40.

In order to give a true representation of a set of data, any scale used on a graph should follow these rules:

- The scale should begin with zero.
- An interval that makes sense for the data that is shown in the graph should be chosen.
- Every interval on the scale must be the same size.

A consumer research company conducted a survey at both Stereo City and Speaker Town. At each store, customers who had just purchased an item were asked if they would return to the store for their next electronics purchase. The results of the survey are shown in the graph.

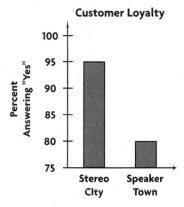

1. What percent of Stereo City's customers said they would make their next electronics purchase there?

   _____ 95% _____

2. What percent of Speaker Town's customers said they would make their next

   electronics purchase there? _____ 80% _____

3. About how many times as high is the bar for Stereo City as the bar for

   Speaker Town? _____ about 4 times as high _____

4. How can you change the graph so that it is not misleading?

   _____ Start the scale at zero and have equal intervals. _____

Reteach RW31

Name _____

LESSON 7.1

Divisibility

The rules for divisibility are:

A number is divisible by:	If:
2	the last digit is 0, 2, 4, 6, or 8.
3	the sum of its digits is divisible by 3.
4	the number formed by the last two digits is divisible by 4.
5	the last digit is 0 or 5.
6	it is divisible by 2 and by 3.
8	the number formed by the last three digits is divisible by 8.
9	the sum of its digits is divisible by 9.
10	the last digit is 0.

To determine whether 3,882 is divisible by 3, follow these steps:

Step 1 Find the sum of the digits of 3,882:

$$3 + 8 + 8 + 2 = 21$$

Step 2 Decide whether the sum, 21, is divisible by 3. Since 3 divides 21 evenly (with no remainder), 21 is divisible by 3. So, 3,882 is *divisible* by 3.

To determine whether 3,882 is divisible by 9, decide whether 21, the sum of the digits, can be divided evenly by 9. Since 21 ÷ 9 = 2 r3, 3,882 is *not divisible* by 9.

To determine whether 7,032 is divisible by 4, follow these steps:

Step 1 Identify the number formed by the last two digits: 32.

Step 2 Decide whether 32 is divisible by 4. Since 4 divides 32 evenly (with no remainder), 32 is *divisible* by 4. So, 7,032 is divisible by 4.

To determine whether 7,032 is divisible by 8, decide whether 032 (the last three digits) is a number divisible by 8. Since 032 ÷ 8 = 4, the number 7,032 is *divisible* by 8.

Determine whether each number is divisible by 2, 3, 4, 5, 6, 8, 9, or 10.

1. 146   2. 369   3. 195   4. 284

   __2__   __3, 9__   __3, 5__   __2, 4__

5. 444   6. 512   7. 788   8. 612

   __2, 3, 4, 6__   __2, 4, 8__   __2, 4__   __2, 3, 4, 6, 9__

9. 2,865   10. 4,470   11. 6,048   12. 3,240

   __3, 5__   __2, 3, 5, 6, 10__   __2, 3, 4, 6, 8, 9__   __2, 3, 4, 5, 6, 8, 9, 10__

RW32   Reteach

Name _____

  LESSON 7.2

Prime Factorization

When you write a composite number as the product of prime factors, you have found the **prime factorization** of the number. A factor tree can help you find the prime factors of a composite number.

What is the prime factorization of 36?

Step 1 Choose any two factors of 36. Draw two lines from 36. Write one factor at the end of each line.

Step 2 Look at the factors. Are they prime? composite? Since they are composite, you must continue to find more factors.

Step 3 Look at the new bottom row of factors. Are they prime? composite? Since they are prime, you have found the prime factors of 36.

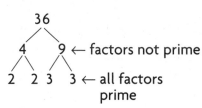

You can write the prime factorization of 36 as $2 \times 2 \times 3 \times 3$, or $2^2 \times 3^2$.

Complete each factor tree to find the prime factors of the number.

1.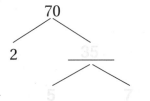

   Prime factorization:

   $2 \times \underline{5} \times \underline{7}$

2.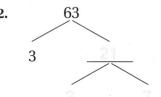

   Prime factorization:

   $3 \times \underline{3} \times \underline{7}$ or $\underline{3^2} \times \underline{7}$

3.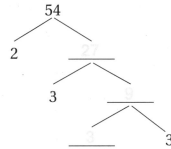

   Prime factorization:

   $2 \times 3 \times \underline{3} \times 3$ or $2 \times \underline{3^3}$

4.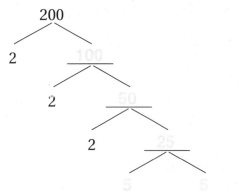

   Prime factorization:

   $2 \times 2 \times 2 \times \underline{5} \times \underline{5}$

   or $\underline{2^3} \times \underline{5^2}$

Reteach RW33

Name _____

LESSON 7.3

Least Common Multiple and Greatest Common Factor

The greatest common factor, or GCF, of two numbers is the largest common factor of both numbers. You can use prime factors to find the GCF of two numbers.

What is the GCF of 28 and 36?

Step 1 Use factor trees to find the prime factors of the numbers.
$28 = 2 \times 2 \times 7$
$36 = 2 \times 2 \times 3 \times 3$

Step 2 Find the prime factors that are in both trees.   2 and 2

Step 3 Multiply the common factors. $2 \times 2 = 4$

Using this method, you discover that the GCF of 28 and 36 is 4.

To find the least common multiple, or LCM, of two numbers, you can list the multiples of each number. The smallest number in both lists is the LCM.

To find the LCM for 8 and 12:

8 → 8, 16, 24, 32, . . .
12 → 12, 24, 36, 48, . . .

The LCM for 8 and 12 is 24.

Complete to find the GCF of 12 and 72. Factor trees may vary.

1. Use factor trees to find the prime factors of the numbers.

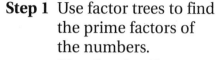

2. Find the common prime factors. __2, 2, and 3__

3. Multiply the common factors. __$2 \times 2 \times 3 = 12$__

4. The GCF of 12 and 72 is __12__.

Find the GCF of each pair of numbers.

5. 6 and 15 __3__   6. 18 and 81 __9__   7. 24 and 84 __12__

Find the LCM of each pair of numbers.

8. 9 and 12 __36__   9. 20 and 15 __60__   10. 18 and 24 __72__

RW34 Reteach

Name _____

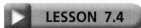

Problem Solving Strategy: Make an Organized List

Organizing data into a list is one way of making sure that you have considered every possibility for a given situation.

Two trains that ran in Great Britain in the 1930s were the Night Scotsman and the Bournemouth Belle. They both ran between London and other cities. The Night Scotsman left London every 4 days and the Bournemouth Belle left every 6 days. If they both left London on Thursday, March 1, on which other dates in March did they leave London on the same day?

Step 1 Think about what you know and what you are asked to find.

- You know how often each train left from London:
  The Night Scotsman left every 4 days.
  The Bournemouth Belle left every 6 days.
  You know that they both left London on Thursday, March 1.
- You are asked to find the other dates in March on which both trains left London on the same day.

Step 2 Plan a strategy to solve.

- Use the strategy *make an organized list*.
- Make a list of the multiples of 4 and the multiples of 6.
- Use the common multiples to determine the other dates in March on which both trains left London on the same day.

Step 3 Solve.

- Carry out the strategy. Make lists of the multiples.

    multiples of 4: 4, 8, **12**, 16, 20, **24**, 28, 32, …

    multiples of 6: 6, **12**, 18, **24**, 30, 36, 42, …

  The common multiples of 4 and 6 that are less than 31 (the number of days in March) are 12 and 24.

So, the trains both left London together 12 days after March 1 and again 24 days after March 1. So, the dates they left on the same day were:

   1 + 12, or March 13   and   1 + 24, or March 25.

Solve the problem by making *an organized list*.

1. A bus company has routes between Chicago and several other cities. Every 5 hours a bus leaves from Chicago for Detroit and every 3 hours another bus leaves for Cleveland. If buses leave for both cities at 6:00 A.M., when is the next time two buses will leave together for these two cities?

   _____9:00 P.M. that same day_____

2. Some buses have seats for 40 people. Other buses can seat only 36. One of each type of bus is filled with families. There are no empty seats and all the families are the same size. What is the greatest number of members that each family can possibly have?

   _____4 members_____

Reteach   RW35

Name _____

LESSON 8.1

Equivalent Fractions and Simplest Form

When the numerator and denominator of a fraction have no common factors other than 1, the fraction is in **simplest form**. You can use a GCF to write a fraction in simplest form.

What is the simplest form of $\frac{32}{56}$?

Step 1 Find the GCF of 32 and 56 by listing the factors of each. The GCF is 8.

32: 1, 2, 4, <u>8</u>, 16, 32
56: 1, 2, 4, 7, <u>8</u>, 14, 28, 56

Step 2 Divide the numerator and denominator by the GCF.

$\frac{32}{56} = \frac{32 \div 8}{56 \div 8} = \frac{4}{7}$

So, $\frac{4}{7}$ is the simplest form of $\frac{32}{56}$.

Complete to find the simplest form of $\frac{88}{104}$.

1. Find the GCF of 88 and 104.   88: ____1, 2, 4, 8, 11, 22, 44, 88____

   The GCF is __8__.   104: ____1, 2, 4, 8, 13, 26, 52, 104____

2. Divide the numerator and denominator by the GCF.   → $\frac{88 \div \boxed{8}}{104 \div \boxed{8}} = \frac{\boxed{11}}{\boxed{13}}$

3. So, __11/13__ is the simplest form of $\frac{88}{104}$.

Complete to find the simplest form of $\frac{78}{120}$.

4. Find the GCF of 78 and 120.   78: ____1, 2, 3, 6, 13, 26, 39, 78____

   The GCF is __6__.   120: __1, 2, 3, 4, 5, 6, 8, 10, 12, 15, 20, 24, 30, 40, 60, 120__

5. Divide the numerator and denominator by the GCF.   → $\frac{78 \div \boxed{6}}{120 \div \boxed{6}} = \frac{\boxed{13}}{\boxed{20}}$

6. So, __13/20__ is the simplest form of $\frac{78}{120}$.

Find the GCF of the pair of numbers. Then write the fraction in simplest form.

7. 4, 10; $\frac{4}{10}$ 8. 8, 12; $\frac{8}{12}$ 9. 18, 36; $\frac{18}{36}$ 10. 21, 60; $\frac{21}{60}$

   2; $\frac{2}{5}$ 4; $\frac{2}{3}$ 18; $\frac{1}{2}$ 3; $\frac{7}{20}$

RW36 Reteach

Mixed Numbers and Fractions

LESSON 8.2

A mixed number is made up of two parts: a whole number and a fraction.
- You can rewrite a mixed number as an equivalent fraction. This type of fraction will have a numerator that is greater than the denominator.
- You can also rewrite a fraction with a numerator greater than the denominator as a mixed number.

Write $4\frac{2}{3}$ as an equivalent fraction.	Write $\frac{19}{5}$ as a mixed number.
Step 1 Multiply the whole number by the denominator of the fraction. $4 \times 3 = 12$	**Step 1** Divide the numerator by the denominator. $19 \div 5 = 3 \text{ r}4$
Step 2 Add the numerator. This sum is the numerator of the equivalent fraction. $12 + 2 = 14$	**Step 2** Use the remainder as the numerator in the fraction part of the mixed number. The denominator is the divisor.
Step 3 Use the same denominator to write the equivalent fraction. $\frac{14}{3}$	$3\frac{4}{5}$ ←remainder ←divisor

Complete.

1. Write $7\frac{3}{5}$ as a fraction.

   Step 1 Multiply the whole number by the __denominator of the fraction__. $7 \times \boxed{5} = \boxed{35}$

   Step 2 Add the __numerator__. This sum is the __numerator__ of the equivalent fraction. $\boxed{35} + \boxed{3} = \boxed{38}$

   Step 3 Use the same __denominator__ to write the equivalent fraction. $\dfrac{\boxed{38}}{\boxed{5}}$

Write the mixed number as a fraction.

2. $5\frac{1}{3}$ __16/3__
3. $4\frac{1}{5}$ __21/5__
4. $5\frac{2}{7}$ __37/7__
5. $6\frac{1}{3}$ __19/3__
6. $7\frac{4}{5}$ __39/5__

Write the fraction as a mixed number.

7. $\frac{17}{3}$ __$5\frac{2}{3}$__
8. $\frac{19}{6}$ __$3\frac{1}{6}$__
9. $\frac{37}{8}$ __$4\frac{5}{8}$__
10. $\frac{52}{9}$ __$5\frac{7}{9}$__
11. $\frac{69}{11}$ __$6\frac{3}{11}$__

Name _____

LESSON 8.3

Compare and Order Fractions

To compare fractions without using a number line, you can use a flowchart approach. Here are the steps for comparing $\frac{1}{3}$ and $\frac{2}{5}$.

Step 1 Are the denominators the same? If YES, then the fraction with the greater numerator is the greater fraction. If NO, go on.	$\frac{1}{3} \quad \frac{2}{5}$ Since the denominators are not the same, go on to the next step.
Step 2 Are the numerators the same? If YES, then the fraction with the smaller denominator is the greater fraction. If NO, go on.	$\frac{1}{3} \quad \frac{2}{5}$ Since the numerators are not the same, go on to the next step.
Step 3 Find the least common denominator of the two fractions.	multiples of 3: 3, 6, 9, 12, **15**, 18, . . . multiples of 5: 5, 10, **15**, 20, 25, . . . 15 is the least common multiple of 3 and 5. So, it is the least common denominator of the fractions.
Step 4 Write fractions equivalent to the original fractions using the common denominator.	$\frac{1}{3} \times \frac{5}{5} = \frac{5}{15} \quad \frac{2}{5} \times \frac{3}{3} = \frac{6}{15}$
Step 5 Compare the fractions using the common denominator.	$\frac{5}{15} < \frac{6}{15}$
Step 6 Write the comparison of the original fractions in the same order.	$\frac{1}{3} < \frac{2}{5}$

Compare the fractions. Write <, >, or = in each ◯.

1. $\frac{4}{7}$ ⓖ $\frac{4}{9}$
2. $\frac{3}{5}$ ⓖ $\frac{2}{5}$
3. $\frac{3}{5}$ ⓛ $\frac{7}{8}$
4. $\frac{8}{10}$ ⓔ $\frac{4}{5}$

5. $\frac{2}{3}$ ⓛ $\frac{5}{6}$
6. $\frac{1}{6}$ ⓛ $\frac{3}{11}$
7. $\frac{6}{15}$ ⓔ $\frac{2}{5}$
8. $\frac{6}{11}$ ⓛ $\frac{5}{9}$

9. $\frac{2}{7}$ ⓛ $\frac{1}{3}$
10. $\frac{3}{5}$ ⓛ $\frac{6}{9}$
11. $\frac{7}{11}$ ⓖ $\frac{5}{12}$
12. $\frac{9}{10}$ ⓖ $\frac{7}{8}$

13. $\frac{2}{5}$ ⓖ $\frac{1}{3}$
14. $\frac{1}{8}$ ⓛ $\frac{2}{11}$
15. $\frac{4}{7}$ ⓔ $\frac{8}{14}$
16. $\frac{9}{14}$ ⓛ $\frac{9}{10}$

RW38 Reteach

Name _____

LESSON 8.5

Fractions, Decimals, and Percents

When you need to change a fraction to a decimal, you can use division.
Write the fraction $\frac{3}{5}$ as a decimal.

Step 1 Set up a division problem, dividing the numerator by the denominator.

$5\overline{)3}$

Step 2 Place a decimal point after the numerator. Write a zero.

$5\overline{)3.0}$

Step 3 Divide as you would with whole numbers.

$\begin{array}{r} 0.6 \\ 5\overline{)3.0} \\ \underline{3\ 0} \\ 0 \end{array}$

So, written as a decimal, $\frac{3}{5} = 0.6$. Recall that 0.6 is a terminating decimal because it ends after the tenths place.

When you need to change a decimal to a fraction, use place value.

Change 0.364 to a fraction.

Step 1 Identify the place value of the last digit in the decimal number.

0.364
↑
thousandths

Step 2 Use the place value of the last digit as the denominator.

$\frac{364}{1,000}$

So, $0.364 = \frac{364}{1,000}$.

Answer the questions to change the fraction to a decimal.

1. $\frac{1}{4}$

   a. What division problem will you use?

   _____ $1 \div 4$ _____

   b. What is the quotient?

   _____ 0.25 _____

2. $\frac{3}{8}$

   a. What division problem will you use?

   _____ $3 \div 8$ _____

   b. What is the quotient?

   _____ 0.375 _____

Use place value to write the decimal as a fraction.

3. 0.6 $\frac{6}{10}$

4. 0.92 $\frac{92}{100}$

5. 0.48 $\frac{48}{100}$

6. 0.137 $\frac{137}{1,000}$

Reteach RW39

Name _____

LESSON 9.1

Estimate Sums and Differences

Some problems need only an estimated answer, not an exact answer. You can estimate with fractions by rounding them to 0, $\frac{1}{2}$, or 1.

Estimate. $\frac{4}{5} + \frac{5}{12}$

Step 1: Look at the first fraction.
- Is 4 much less than 5? No.
- Is 4 about half of 5? No.
- Is 4 almost the same as 5? Yes.
- Since the numerator is almost the same as the denominator, round the fraction to 1.

Step 2: Look at the second fraction.
- Is 5 almost the same as 12? No.
- Is 5 about half of 12? Yes.
- Since the numerator is about half of the denominator, round the fraction to $\frac{1}{2}$.

Step 3: Add the rounded fractions to get an estimated answer. $1 + \frac{1}{2} = 1\frac{1}{2}$

Follow the same steps to estimate a difference.

Estimate. $\frac{8}{9} - \frac{1}{10}$

Step 1: Look at the first fraction.
- Is 8 much less than 9? No.
- Is 8 about half of 9? No.
- Is 8 almost the same as 9? Yes.
- So, round the fraction to 1.

Step 2: Look at the second fraction.
- Is 1 much less than 10? Yes.
- So, round the fraction to 0.

Step 3: Find the difference between the rounded fractions to get an estimated answer. $1 - 0 = 1$

Estimate the sum or difference. Possible estimates are given.

1. $\frac{6}{7} - \frac{1}{3}$ about $\frac{1}{2}$

2. $\frac{11}{12} + \frac{4}{7}$ about $1\frac{1}{2}$

3. $\frac{1}{9} + \frac{5}{8}$ about $\frac{1}{2}$

4. $\frac{8}{11} - \frac{1}{15}$ about $\frac{1}{2}$

5. $\frac{13}{14} - \frac{11}{12}$ about 0

6. $\frac{5}{6} + \frac{1}{7}$ about 1

7. $\frac{7}{8} - \frac{4}{9}$ about $\frac{1}{2}$

8. $\frac{3}{5} + \frac{1}{2}$ about 1

9. $\frac{11}{12} - \frac{15}{16}$ about 0

10. $\frac{3}{7} - \frac{1}{8}$ about $\frac{1}{2}$

RW40 Reteach

Name _____

LESSON 9.3

Add and Subtract Fractions

To add or subtract unlike fractions, first change them to equivalent fractions with the same denominator.

Find the sum. $\frac{1}{4} + \frac{2}{3}$

Step 1: Find the LCM of the denominators. The LCM of 4 and 3 is 12.
So, the LCD of $\frac{1}{4}$ and $\frac{2}{3}$ is 12.

Step 2: Multiply to write equivalent fractions, using the LCD.

$$\frac{1}{4} = \frac{1 \times 3}{4 \times 3} = \frac{3}{12}$$

$$\frac{2}{3} = \frac{2 \times 4}{3 \times 4} = \frac{8}{12}$$

Step 3: Add the numerators. Write the sum over the denominator.

$$\frac{3}{12} + \frac{8}{12} = \frac{11}{12}$$ ← Remember, keep the denominator the same.

Step 4: Write the answer as a fraction or as a mixed number in simplest form.

$\frac{11}{12}$ is already in simplest form.

$\frac{1}{4} + \frac{2}{3} = \frac{11}{12}$

Follow the same steps to subtract unlike fractions.

Complete to find each sum. Write the answer in simplest form.

1. Find. $\frac{2}{5} + \frac{3}{10}$

The LCM of 5 and 10 is 10.

So, $\frac{2}{5} + \frac{3}{10} = \frac{4}{10} + \frac{3}{10} = \frac{7}{10}$.

2. Find. $\frac{7}{8} + \frac{2}{3}$

The LCM of 8 and 3 is 24.

So, $\frac{7}{8} + \frac{2}{3} = \frac{21}{24} + \frac{16}{24} = \frac{37}{24} = 1\frac{13}{24}$.

3. Find. $\frac{1}{4} + \frac{5}{6}$

The LCM of 4 and 6 is 12.

So, $\frac{1}{4} + \frac{5}{6} = \frac{3}{12} + \frac{10}{12} = \frac{13}{12} = 1\frac{1}{12}$.

4. Find. $\frac{1}{6} + \frac{3}{8}$

The LCM of 6 and 8 is 24.

So, $\frac{1}{6} + \frac{3}{8} = \frac{4}{24} + \frac{9}{24} = \frac{13}{24}$.

Find the sum or difference. Write the answer in simplest form.

5. $\frac{2}{7} + \frac{1}{2}$

$\frac{11}{14}$

6. $\frac{11}{12} - \frac{7}{8}$

$\frac{1}{24}$

7. $\frac{4}{5} + \frac{1}{3}$

$\frac{17}{15}$ or $1\frac{2}{15}$

8. $\frac{3}{4} - \frac{1}{5}$

$\frac{11}{20}$

Reteach RW41

Name _____

LESSON 9.4

Add and Subtract Mixed Numbers

Zack is working on a science project. He needs a $1\frac{1}{8}$-ft piece of wire and a $2\frac{1}{4}$-ft piece of wire. Zack wants to know the total amount of wire he needs for the project. He decides to make a diagram to find the total.

Step 1 Zack draws a diagram that represents each piece of wire.

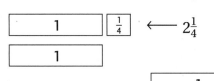

Step 2 Zack combines the whole numbers. He draws equivalent fractions with the LCD of 8 to combine the fractions.

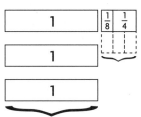

$1 + 2 = 3$ $\frac{1}{8} + \frac{2}{8} = \frac{3}{8}$

Step 3 Then Zack adds the whole numbers and the fractions.

$3 + \frac{3}{8} = 3\frac{3}{8}$

So, Zack needs $3\frac{3}{8}$ ft of wire.

Draw a diagram to find each sum or difference. Check students' drawings.

1. $3\frac{1}{4} + 2\frac{1}{2} =$ _____ $5\frac{3}{4}$

2. $2\frac{5}{6} - 1\frac{1}{3} =$ _____ $1\frac{1}{2}$

3. $1\frac{3}{4} + 2\frac{1}{8} =$ _____ $3\frac{7}{8}$

4. $3\frac{4}{5} - 1\frac{1}{2} =$ _____ $2\frac{3}{10}$

RW42 Reteach

Name _____

LESSON 9.6

Subtract Mixed Numbers

Mrs. Ruiz buys $4\frac{1}{2}$ lb of apples. She uses $1\frac{2}{3}$ lb to bake apple tarts. How many pounds of apples are left?

Step 1 The LCD of $\frac{1}{2}$ and $\frac{2}{3}$ is 6.
Rename the fractions using the LCD.

$$4\frac{1}{2} = 4\frac{3}{6}$$
$$-1\frac{2}{3} = 1\frac{4}{6}$$

Step 2 Since you can't subtract $\frac{4}{6}$ from $\frac{3}{6}$, rename $4\frac{3}{6}$.
Think: $4\frac{3}{6} = 3 + \frac{6}{6} + \frac{3}{6} = 3\frac{9}{6}$

Now, subtract the fractions. Then subtract the whole numbers.

$$4\frac{1}{2} = 4\frac{3}{6} = 3\frac{9}{6}$$
$$-1\frac{2}{3} = 1\frac{4}{6} = 1\frac{4}{6}$$
$$2\frac{5}{6}$$

So, $4\frac{1}{2} - 1\frac{2}{3} = 2\frac{5}{6}$

Mrs. Ruiz has $2\frac{5}{6}$ lbs of apples left.

Find the difference. Write the answer in simplest form.

1. $6\frac{2}{5}$
   $-3\frac{7}{10}$

2. $4\frac{1}{6}$
   $-2\frac{3}{4}$

3. $9\frac{3}{7}$
   $-5\frac{1}{2}$

4. $11\frac{1}{3}$
   $-7\frac{8}{9}$

5. $10\frac{3}{8}$
   $-3\frac{1}{2}$

6. $6\frac{1}{10}$
   $-4\frac{4}{5}$

7. $9\frac{1}{8}$
   $-4\frac{3}{4}$

8. $12\frac{2}{5}$
   $-6\frac{1}{2}$

9. $8\frac{2}{3}$
   $-5\frac{5}{6}$

Reteach **RW43**

Name _____

LESSON 9.7

Problem Solving Strategy: Draw a Diagram

Some problems become clearer when you draw a diagram to represent the information you are given.

Ms. Kaminsky is teaching her class a game in which students stand in a circle. Two girls stand together, then 1 boy, then 2 girls, then 1 boy, and so on. Altogether, 7 boys are standing in the circle along with all the girls in the class. If there are 27 students in the class, how many boys are not participating?

Step 1: Think about what you know and what you are asked to find.
- You know that all the girls in the class are in the circle.
- You know that there are 7 boys in the circle.
- You know that for every 2 girls in the circle, there is 1 boy.
- You are asked to find how many boys in the class are not in the circle.

Step 2: Plan a strategy to solve.
- Use the strategy *draw a diagram*.
- Draw what you know from the problem.

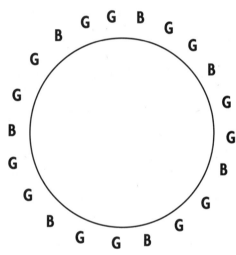

Step 3: Solve.
- Carry out the strategy.

The diagram shows that if 7 boys are in the circle, then there must be 14 girls.

$7 + 14 = 21$ and $27 - 21 = 6$.

So, 6 boys in the class are not participating.

Use the strategy *draw a diagram* to help you solve.

1. In another game, the arrangement around the circle is reversed. Two boys stand together, then 1 girl, then 2 boys, then 1 girl, and so on. In this version of the game, what is the greatest number of students in Ms. Kaminsky's class that can participate at one time?

2. The desks are arranged in rows of 6 from the front of the room to the back. In all, there are 5 rows. Arthur sits in the fourth seat of the first row. Andrea sits 3 rows over and 1 seat back. Charles sits 3 seats in front of Andrea. Juan sits 2 rows over and 2 seats farther back than Charles. Where is Juan compared to Arthur?

18 students: 12 boys and 6 girls   _next to Arthur_

RW44 Reteach

Name _____

LESSON 10.1

Estimate Products and Quotients

You can often estimate the product or quotient of two mixed numbers by rounding each of them to the nearest whole number. Using a number line may help you round in the appropriate direction.

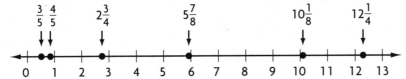

Use rounding to multiply $2\frac{3}{4} \times 10\frac{1}{8}$

Step 1 Round each mixed number to the nearest whole number.

$$2\frac{3}{4} \times 10\frac{1}{8}$$
$$\downarrow \quad \downarrow$$
$$3 \times 10$$

Step 2 Multiply the rounded values.

$$3 \times 10 = 30$$

The product is about 30.

Use rounding to divide $12\frac{1}{4} \div 5\frac{7}{8}$

Step 1 Round each mixed number to the nearest whole number.

$$12\frac{1}{4} \div 5\frac{7}{8}$$
$$\downarrow \quad \downarrow$$
$$12 \div 6$$

Step 2 Divide the rounded values.

$$12 \div 6 = 2$$

The quotient is about 2.

When a fraction is between 0 and 1, round it to 0, $\frac{1}{2}$, or 1, whichever is closest. Remember that you cannot divide by 0.

Use rounding to multiply $\frac{4}{5} \times 10\frac{1}{8}$

$$\frac{4}{5} \times 10\frac{1}{8} \rightarrow 1 \times 10$$

The product is about 1×10, or 10.

Use rounding to divide $12\frac{1}{4} \div \frac{3}{5}$

$$12\frac{1}{4} \div \frac{3}{5} \rightarrow 12 \div \frac{1}{2}$$

The quotient is about $12 \div \frac{1}{2} = 12 \times 2$, or 24.

Complete the estimation of each product or quotient. Possible answers are given.

1. Estimate $15\frac{1}{8} \times 1\frac{7}{8}$

   $15\frac{1}{8} \times 1\frac{7}{8} \rightarrow$ __15__ $\times$ __2__

   $=$ __30__

2. Estimate $24\frac{5}{6} \div 4\frac{4}{5}$

   $24\frac{5}{6} \div 4\frac{4}{5} \rightarrow$ __25__ $\div$ __5__

   $=$ __5__

3. Estimate $\frac{3}{5} \times 48$

   $\frac{3}{5} \times 48 \rightarrow$ __2__ $\times$ __48__

   $=$ __24__

4. Estimate $59\frac{7}{8} \div 15\frac{1}{8}$

   $59\frac{7}{8} \div 15\frac{1}{8} \rightarrow$ __60__ $\div$ __15__

   $=$ __4__

Reteach RW45

Name _____

LESSON 10.2

Multiply Fractions

Bo knows that $\frac{2}{3}$ of the students in his class play soccer. Of those students, $\frac{1}{6}$ are in the school band. Bo wants to know what fraction of his class play soccer and are in the school band.

Step 1 Write a multiplication sentence. $\qquad \frac{2}{3} \times \frac{1}{6} = \blacksquare$

Step 2 Multiply the numerators. $\qquad \frac{2 \times 1}{3 \times 6} = \frac{2}{18}$
Multiply the denominators.

Step 3 Divide the numerator and $\qquad = \frac{2 \div 2}{18 \div 2}$
denominator by the GCF, 2.

Step 4 Write the product in simplest form. $\qquad = \frac{1}{9}$

So, $\frac{1}{9}$ of Bo's class play soccer and are in the school band.

Complete to find each product.

1. $\frac{3}{4} \times \frac{2}{9} = \dfrac{3 \times \boxed{}}{\boxed{} \times 9}$

    $= \dfrac{6}{\boxed{}}$

    $= \dfrac{6 \div \boxed{}}{\boxed{} \div \boxed{}}$

    $= \dfrac{\boxed{}}{\boxed{}}$

2. $\frac{1}{8} \times \frac{2}{3} = \dfrac{1 \times 2}{8 \times \boxed{}}$

    $= \dfrac{\boxed{}}{24}$

    $= \dfrac{\boxed{} \div \boxed{}}{24 \div \boxed{}}$

    $= \dfrac{\boxed{}}{\boxed{}}$

Use GCFs to simplify the fractions. Write the product in simplest form.

3. $\frac{5}{6} \times \frac{2}{3}$ _____

4. $\frac{3}{4} \times \frac{5}{12}$ _____

5. $\frac{5}{9} \times \frac{3}{10}$ _____

6. $\frac{3}{4} \times \frac{6}{7}$ _____

7. $\frac{4}{5} \times \frac{5}{8}$ _____

8. $\frac{5}{6} \times \frac{12}{25}$ _____

9. $\frac{2}{3} \times \frac{9}{20}$ _____

10. $\frac{3}{8} \times \frac{4}{15}$ _____

RW46 Reteach

Name _____

LESSON 10.3

Multiply Mixed Numbers

Barbara bought $2\frac{1}{4}$ dozen donuts. Her family ate $\frac{2}{3}$ of them. How many dozen did her family eat?

Step 1 What is $\frac{2}{3}$ of $2\frac{1}{4}$? $\frac{2}{3} \times 2\frac{1}{4} = \blacksquare$

Find $\frac{2}{3} \times 2\frac{1}{4}$.

Step 2 Write the mixed number as a fraction. $\frac{2}{3} \times 2\frac{1}{4} = \frac{2}{3} \times \frac{9}{4}$

Step 3 Use the GCF to simplify.
The GCF of 2 and 4 is 2.
The GCF of 3 and 9 is 3.

$$= \frac{\overset{1}{2}}{\underset{1}{3}} \times \frac{\overset{3}{9}}{\underset{2}{4}}$$

Step 4 Multiply. $= \frac{1}{1} \times \frac{3}{2} = \frac{3}{2} = 1\frac{1}{2}$

So, Barbara's family ate $1\frac{1}{2}$ dozen.

Find the product. Write it in simplest form.

1. $\frac{2}{5} \times 1\frac{2}{3}$

2. $4\frac{1}{2} \times 6\frac{2}{3}$

3. $2\frac{1}{5} \times 3\frac{1}{8}$

4. $\frac{8}{9} \times 3\frac{3}{4}$

5. $1\frac{3}{7} \times 1\frac{5}{9}$

6. $2\frac{1}{3} \times 1\frac{5}{7}$

7. $\frac{5}{8} \times 4\frac{4}{5}$

8. $2\frac{3}{4} \times 1\frac{3}{5}$

9. $1\frac{1}{3} \times 5\frac{3}{4}$

10. $2\frac{4}{9} \times 1\frac{1}{2}$

11. $1\frac{2}{5} \times \frac{5}{8}$

12. $3\frac{5}{6} \times 2\frac{1}{4}$

13. $1\frac{1}{7} \times \frac{9}{10}$

14. $\frac{4}{7} \times 2\frac{3}{8}$

15. $\frac{5}{7} \times 4\frac{1}{5}$

Reteach RW47

Name _____

LESSON 10.5

Divide Fractions and Mixed Numbers

Beth is working on a science project. She needs $\frac{2}{3}$-yd pieces of wire for the project. She bought a 6-yd piece of wire at the hardware store. How many $\frac{2}{3}$-yd pieces can she cut from this piece?

Step 1 Write a division sentence to find this amount.

$$\frac{6}{1} \div \frac{2}{3} = \blacksquare$$

Step 2 Use the reciprocal of the divisor to write a multiplication problem.

$$\frac{6}{1} \div \frac{2}{3} = \frac{6}{1} \times \frac{3}{2}$$

Think: the reciprocal of $\frac{2}{3}$ is $\frac{3}{2}$.

Step 3 Simplify.

$$= \frac{\overset{3}{\cancel{6}}}{1} \times \frac{3}{\underset{1}{\cancel{2}}}$$

Step 4 Multiply.

$$= \frac{3}{1} \times \frac{3}{1} = \frac{9}{1} = 9$$

So, Beth can cut 9 pieces of wire.

Find the quotient. Write it in simplest form.

1. $8 \div \frac{3}{4}$

    $10\frac{2}{3}$

2. $\frac{5}{9} \div \frac{2}{3}$

    $\frac{5}{6}$

3. $2\frac{4}{5} \div \frac{2}{3}$

    $4\frac{1}{5}$

4. $2\frac{1}{10} \div \frac{3}{5}$

    $3\frac{1}{2}$

5. $12 \div \frac{4}{5}$

    15

6. $2\frac{5}{8} \div \frac{3}{4}$

    $3\frac{1}{2}$

7. $4\frac{2}{5} \div 1\frac{2}{3}$

    $2\frac{16}{25}$

8. $2\frac{1}{5} \div 1\frac{3}{10}$

    $1\frac{9}{13}$

9. $3\frac{2}{7} \div 1\frac{6}{7}$

    $1\frac{10}{13}$

10. $5\frac{5}{6} \div 3\frac{1}{3}$

    $1\frac{3}{4}$

11. $6\frac{1}{2} \div 2\frac{3}{4}$

    $2\frac{4}{11}$

12. $\frac{5}{12} \div \frac{5}{8}$

    $\frac{2}{3}$

13. $27 \div \frac{3}{8}$

    72

14. $\frac{4}{7} \div \frac{1}{2}$

    $1\frac{1}{7}$

15. $\frac{7}{8} \div \frac{1}{4}$

    $3\frac{1}{2}$

RW48 Reteach

Name _____

LESSON 10.6

Problem Solving Skill: Choose the Operation

It is often helpful to think about the kind of answer you need to solve a problem before you decide which operation or operations to use. Here are some different problem types and the operations used to solve them.

- **Add to combine two or more like measures, such as length or weight.**

  Lolly put $3\frac{1}{2}$ lb of sugar into a can containing 6 lb of sugar. The total amount of sugar in the can is $3\frac{1}{2} + 6$, or $9\frac{1}{2}$ lb.

- **Subtract to take away an amount or to compare like measures.**

  Perry spilled $2\frac{1}{3}$ lb of flour out of a 10-lb bag. The flour remaining in the bag weighs $10 - 2\frac{1}{3}$, or $7\frac{2}{3}$ lb. There is $7\frac{2}{3} - 2\frac{1}{3}$, or $5\frac{1}{3}$ lb more flour in the bag than was spilled.

- **Multiply to combine a number of equal measures or to calculate a new type of measure.**

  A square $3\frac{1}{2}$ yd on each side has a perimeter of $4 \times 3\frac{1}{2}$, or 14 yd. The square also has an area of $3\frac{1}{2} \times 3\frac{1}{2}$, or $12\frac{1}{4}$ yd².

- **Divide to determine how many parts of equal size are in a measure or to determine the size of several equal parts.**

  A $9\frac{1}{2}$ in. long board is cut into 4 equal lengths. Each piece has a length of $9\frac{1}{2} \div 4$, or $2\frac{3}{8}$ in. If a 6 ft board is cut in $1\frac{1}{2}$ ft lengths, there will be $6 \div 1\frac{1}{2}$, or 4 pieces.

Name the operation you would use to solve the problem. Then solve it.

1. Orange juice comes in 1-gal (128-oz) containers. William uses $12\frac{1}{2}$ oz of orange juice to make his favorite fruit smoothie. How many smoothies should he be able to make from one container of orange juice?

   divide: $128 \div 12\frac{1}{2}$;

   about 10 smoothies

2. William spills about $3\frac{1}{2}$ oz of liquid each time he makes a smoothie. He makes an average of 44 smoothies each day. How many ounces of liquid would he spill on a typical day?

   multiply: $3\frac{1}{2} \times 44$;

   about 154 oz

Reteach RW49

Name _____

LESSON 10.7

Algebra: Fraction Expressions and Equations

Algebraic expressions can have fractions in them. Also, fractional values can replace a variable in the expression.

Let's explore the fraction expressions $x + \frac{2}{5}$ and $\frac{3}{4}y$ for several values of the variables x and y.

x	$x + \frac{2}{5}$
$\frac{1}{5}$	$\frac{1}{5} + \frac{2}{5} = \frac{3}{5}$
1	$1 + \frac{2}{5} = 1\frac{2}{5}$
10	$10 + \frac{2}{5} = 10\frac{2}{5}$

y	$\frac{3}{4}y$
$\frac{1}{2}$	$\frac{3}{4} \times \frac{1}{2} = \frac{3}{8}$
2	$\frac{3}{4} \times 2 = 1\frac{1}{2}$
10	$\frac{3}{4} \times 10 = 7\frac{1}{2}$

Notice that the expression $x + \frac{2}{5}$ adds $\frac{2}{5}$ to any value of x, and that the expression $\frac{3}{4}y$ multiplies any value of y by $\frac{3}{4}$.

When solving an equation involving fractions and fractional expressions, often you can use your number sense to decide what value for the variable makes the equation true.

- Solve the equation $x + \frac{2}{5} = \frac{3}{5}$.

  Ask yourself: *What number can I add to $\frac{2}{5}$ to get $\frac{3}{5}$?*

  You can add $\frac{1}{5}$ to $\frac{2}{5}$ to get $\frac{3}{5}$. So, $x = \frac{1}{5}$.

- Solve the equation $\frac{3}{4}y = \frac{30}{4}$.

  Ask yourself: *What can I multiply $\frac{3}{4}$ by to get $\frac{30}{4}$?*

  You can multiply $\frac{3}{4}$ by 10 to get $\frac{30}{4}$. So, $y = 10$.

Evaluate the expression.

1. $x + \frac{1}{10}$ for $x = \frac{1}{2}$ _____
2. $\frac{1}{8}y$ for $y = \frac{4}{9}$ _____
3. $\frac{1}{3} - z$ for $z = \frac{1}{4}$ _____

Use mental math to solve the equation.

4. $y + \frac{1}{2} = \frac{5}{2}$ $y = 2$
5. $\frac{1}{10}m = 2$ $m = 20$
6. $\frac{1}{3}t = 1$ $t = 3$

7. $\frac{1}{4}x = 2$ $x = 8$
8. $n + \frac{3}{8} = \frac{1}{2}$ $n = \frac{1}{8}$
9. $w - \frac{3}{10} = \frac{1}{5}$ $w = \frac{1}{2}$

10. $\frac{5}{6} - x = \frac{3}{4}$ $x = \frac{1}{12}$
11. $t - \frac{1}{3} = 1$ $t = \frac{4}{3}$
12. $a + \frac{1}{3} = 7$ $a = 6\frac{2}{3}$

Name _____

LESSON 11.1

Understand Integers

A diver started out at the bottom of the ocean, 250 feet below sea level. He came to the surface and then climbed a hill 300 feet above sea level. How can you represent these numbers?

Integers are numbers that can show opposite directions.

250 feet below sea level is $^-250$ ft.   300 feet above sea level is $^+300$ ft.

Every integer, except zero, has an opposite.

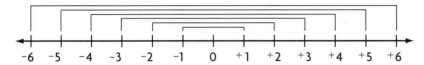

The absolute value of a number is used to show how far a number is from 0.

Find $|^-2|$: This is read as the absolute value of negative 2.

So, $|^-2| = 2$.

Write the opposite integer.

1. $^-8$ 2. $^+7$ 3. $^+11$ 4. $^-9$ 5. $^-14$ 6. $^+17$

7. $^-3$ 8. $^+12$ 9. $^+23$ 10. $^-30$ 11. $^+33$ 12. $^-50$

Find the absolute value.

13. $|^-6|$ 14. $|^+6|$ 15. $|^-8|$ 16. $|^-21|$ 17. $|^+13|$ 18. $|^-26|$

19. $|^-45|$ 20. $|^+56|$ 21. $|^-77|$ 22. $|^+92|$ 23. $|^-345|$ 24. $|^+880|$

Reteach   RW51

Rational Numbers

Between any two rational numbers, you can always find other rational numbers.

Fractions or Mixed Numbers

Find a rational number between $3\frac{3}{4}$ and $3\frac{7}{8}$.

- Use a common denominator to write equivalent fractions. 8 is a common denominator of 4 and 8.

$$\frac{3}{4} = \frac{6}{8} \qquad \frac{7}{8}$$

- There are no eighths between $\frac{6}{8}$ and $\frac{7}{8}$.
Use a greater common denominator. Try 16.

$$\frac{3}{4} = \frac{12}{16} \qquad \frac{7}{8} = \frac{14}{16}$$

$$3\frac{3}{4} = 3\frac{12}{16} \qquad 3\frac{7}{8} = 3\frac{14}{16}$$

$3\frac{13}{16}$ is between $3\frac{12}{16}$ and $3\frac{14}{16}$.

So, $3\frac{13}{16}$ is between $3\frac{3}{4}$ and $3\frac{7}{8}$.

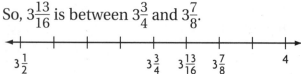

Decimals

Find a rational number between 1.3 and 1.4.

- Add a zero to each number. $1.3 = 1.30 \qquad 1.4 = 1.40$

- Find a number between the two numbers.   1.31, 1.32, 1.33, 1.34, 1.35, 1.36, 1.37, 1.38, and 1.39 are between 1.30 and 1.40.

So, 1.31, 1.33, 1.36, and 1.38 are some of the numbers between 1.3 and 1.4.

Find a rational number between the two given numbers. Possible answers are given.

1. $\frac{1}{2}$ and $\frac{3}{4}$ 2. $\frac{1}{8}$ and $\frac{1}{4}$ 3. $1\frac{1}{3}$ and $1\frac{2}{3}$ 4. $-3\frac{1}{2}$ and $-3\frac{3}{4}$

    $\frac{5}{8}$ $\frac{3}{16}$ $1\frac{1}{2}$ $-3\frac{5}{8}$

5. 3.6 and 3.7   6. 1.2 and 1.3   7. -5.9 and -6 8. 2.11 and 2.12

    3.61   1.22 -5.95 2.111

9. $5\frac{1}{4}$ and $5\frac{3}{4}$ 10. -2.75 and -2.76 11. $\frac{7}{10}$ and 0.8 12. $1\frac{1}{10}$ and 1.2

    $5\frac{1}{2}$ -2.757 $\frac{3}{4}$, or 0.75 $\frac{23}{20}$, or 1.15

Name _____

LESSON 11.3

Compare and Order Rational Numbers

You can compare rational numbers in decimal or fraction form.

Using Decimals

Compare $\frac{7}{25}$ and 0.35.

- Write the number that is not a decimal in decimal form.

$$\frac{7}{25} = 25\overline{)7.00} = 0.28$$

- Compare decimals using place value.

$0.28 < 0.35$

So, $\frac{7}{25} < 0.35$.

Using Fractions

Compare $\frac{1}{4}$ and 0.3.

- Write the number that is not a fraction in fraction form.

$0.3 = \frac{3}{10}$

- Rewrite the fractions with the same denominator. Use 20 as a common denominator.

$\frac{1}{4} = \frac{5}{20}$ $\frac{3}{10} = \frac{6}{20}$

- Compare the two fractions.

$\frac{5}{20} < \frac{6}{20}$

So, $\frac{1}{4} < 0.3$.

Compare. Write < or >.

1. $\frac{2}{5}$ __>__ 0.2
2. 0.65 __<__ $\frac{2}{3}$
3. 8.9 __>__ $8\frac{4}{5}$
4. $^-4\frac{1}{8}$ __>__ $^-4.3$

Compare the rational numbers and order them from least to greatest.

5. 4.2, 2.4, $\frac{8}{2}$, $\frac{2}{5}$

$\frac{2}{5} < 2.4 < \frac{8}{2} < 4.2$

6. $\frac{7}{5}$, $\frac{2}{3}$, 0.2, 0.8

$0.2 < \frac{2}{3} < 0.8 < \frac{7}{5}$

7. 0.1, 0.6, 0.9, 0, $\frac{1}{9}$

$0 < 0.1 < \frac{1}{9} < 0.6 < 0.9$

8. $\frac{1}{3}$, $\frac{1}{5}$, $\frac{1}{8}$, 0.1

$0.1 < \frac{1}{8} < \frac{1}{5} < \frac{1}{3}$

9. $^-1.4$, $^-1.5$, 1.2, $\frac{6}{4}$

$^-1.5 < ^-1.4 < 1.2 < \frac{6}{4}$

10. 2.1, $^-3.8$, $\frac{10}{2}$, $\frac{^-3}{4}$

$^-3.8 < \frac{^-3}{4} < 2.1 < \frac{10}{2}$

11. 5.8, 4.9, 5.7, $2\frac{3}{5}$, 0.58

$0.58 < 2\frac{3}{5} < 4.9 < 5.7 < 5.8$

12. $^-5.6$, $^-4.62$, 2.34, $^-4.68$

$^-5.6 < ^-4.68 < ^-4.62 < 2.34$

Reteach   RW53

Name _____

Problem Solving Strategy: Use Logical Reasoning

When a problem presents a lot of information, logical reasoning can often be used in combination with organizing the information in a table.

Chris, Keiko, Rosa, and Jamal are officers in their school's student government. One of them is president, one is vice president, one is secretary, and one is treasurer. Rosa is the president. Chris is not the treasurer. Keiko is the vice president. What office does Jamal hold?

Step 1: Think about what you know and what you are asked to find.

You know: the names of the officers and the offices;

Rosa is the president, Chris is not the treasurer, and Keiko is the vice president.

You are asked to find the office held by Jamal.

Step 2: Plan a strategy to solve.
- Use the strategy *use logical reasoning*.
- Organize the information in a table, using one at a time.

Step 3: Solve.
- Since Rosa is president, put a Y for *yes* in the box where the Rosa column and the President row cross. Put an N for *no* in the other boxes in this row and column, since Rosa cannot hold any other office and nobody else can be president.
- Since Chris is not the treasurer, put an N where the Chris column and the Treasurer row cross.
- Since Keiko is the vice president, put a Y where the Keiko column and the Vice President row cross. Put Ns in the other boxes in this row and column.
- The only office left for Chris is secretary. Put a Y where the Chris column and the Secretary row cross.
- Therefore, Jamal must be the treasurer.

	Chris	Keiko	Rosa	Jamal
President	N	N	Y	N
Vice President	N	Y	N	N
Secretary	Y	N	N	N
Treasurer	N	N	N	Y

Solve the problem using logical reasoning.

1. The Raiders, Rangers, Cougars, and Lions are in one division of a baseball league. Currently, the Raiders are ahead of the Rangers and Cougars. The Raiders and Cougars are behind the Lions. The Cougars are in last place. Who is in first place?

   _____ Lions _____

2. José, Cody, and Sid are students. One of them is in sixth grade, one is in seventh, and one is in eighth. The seventh-grader and José walk to school together. Sid plays ball with the eighth-grader. Cody is in sixth grade. Who is in seventh grade?

   _____ Sid _____

Name _____

LESSON 12.2

Add Integers

You can use a number line to help you add integers.
Find the sum $^{+}4 + {}^{-}8$.

- Draw a number line.

- Start at 0. Move 4 spaces to the right to show $^{+}4$.

- From $^{+}4$, move 8 spaces to the left to show $^{-}8$.
So, $^{+}4 + {}^{-}8 = {}^{-}4$.

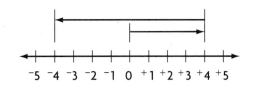

Find the sum $^{-}3 + {}^{-}5$.
- Draw a number line.

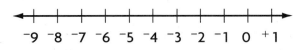

- Start at 0. Move 3 spaces to the left to show $^{-}3$.

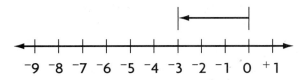

- From $^{-}3$, move 5 spaces to the left to show $^{-}5$.
So, $^{-}3 + {}^{-}5 = {}^{-}8$.

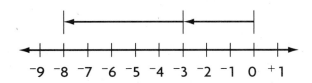

Use a number line to find each sum. Check students' number lines.

1. $^{+}7 + {}^{-}4 = $ ___$^{+}3$___
2. $^{-}8 + {}^{-}2 = $ ___$^{-}10$___
3. $^{+}5 + {}^{-}3 = $ ___$^{+}2$___

4. $^{-}6 + {}^{-}3 = $ ___$^{-}9$___
5. $^{+}5 + {}^{-}8 = $ ___$^{-}3$___
6. $^{-}4 + {}^{-}9 = $ ___$^{-}13$___

7. $^{+}7 + {}^{-}2 = $ ___$^{+}5$___
8. $^{-}3 + {}^{+}5 = $ ___$^{+}2$___
9. $^{-}1 + {}^{-}6 = $ ___$^{-}7$___

10. $^{-}2 + {}^{+}6 = $ ___$^{+}4$___
11. $^{+}8 + {}^{-}3 = $ ___$^{+}5$___
12. $^{-}2 + {}^{-}5 = $ ___$^{-}7$___

Reteach RW55

Name _____

LESSON 12.4

Subtract Integers

In New York City, the 9:00 A.M. temperature reading was ⁻5°C. By noon, the temperature had dropped 3°C. What was the temperature reading at noon?

Find ⁻5 − ⁺3.

You can find the difference of two integers by adding the opposite of the integer you are subtracting.

The opposite of ⁺3 is ⁻3.
⁻5 − ⁺3 becomes ⁻5 + ⁻3.

- Draw a number line.

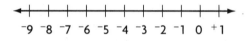

- Start at 0. Move 5 spaces to the left to show ⁻5.

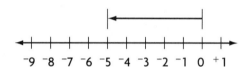

- From ⁻5, move 3 spaces to the left to add ⁻3.

So, at noon the temperature was ⁻8°C.

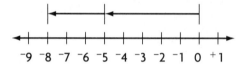

Rewrite the subtraction problem as an addition problem.

1. ⁻4 − ⁻3   2. ⁺8 − ⁻8   3. ⁺5 − ⁺9   4. ⁻6 − ⁺4

   ⁻4 + ⁺3   ⁺8 + ⁺8   ⁺5 + ⁻9   ⁻6 + ⁻4

Use a number line to find the difference. Check students' number lines.

5. ⁺6 − ⁺9 = ⁻3   6. ⁻5 − ⁻4 = ⁻1   7. ⁺7 − ⁻5 = ⁺12

8. ⁻8 − ⁺6 = ⁻14   9. ⁺6 − ⁺4 = ⁺2   10. ⁻9 − ⁺5 = ⁻14

11. ⁻3 − ⁻7 = ⁺4   12. ⁺1 − ⁺8 = ⁻7   13. ⁺2 − ⁻7 = ⁺9

14. ⁻7 − ⁺5 = ⁻12   15. ⁺9 − ⁺6 = ⁺3   16. ⁻6 − ⁺4 = ⁻10

17. ⁻4 − ⁻8 = ⁺4   18. ⁺2 − ⁺9 = ⁻7   19. ⁺4 − ⁻3 = ⁺7

RW56 Reteach

LESSON 12.5

Name _____

Multiply and Divide Integers

You can use patterns to multiply integers. Look at the patterns.

$3 \times 2 = 6$	$^-2 \times 3 = ^-6$
$3 \times 1 = 3$	$^-2 \times 2 = ^-4$
$3 \times 0 = 0$	$^-2 \times 0 = 0$
$3 \times ^-1 = ^-3$	$^-2 \times ^-2 = 4$
$3 \times ^-2 = ^-6$	$^-2 \times ^-3 = 6$

When you multiply a positive integer by a negative integer, the product is a negative integer:

$$8 \times ^-2 = ^-16 \text{ and } ^-8 \times 2 = ^-16$$

When you multiply two negative integers, the product is a positive integer.

$$^-6 \times ^-5 = 30$$

You can use related multiplication problems to divide integers.

$3 \times 2 = 6$	$6 \div 3 = 2$
$^-3 \times 2 = ^-6$	$^-6 \div ^-3 = 2$
$3 \times ^-2 = ^-6$	$^-6 \div 3 = ^-2$
$^-3 \times ^-2 = 6$	$6 \div ^-3 = ^-2$

When you divide two negative integers, the quotient is positive.

$$^-12 \div ^-2 = 6$$

When you divide a positive integer by a negative integer, the quotient is negative.

$$12 \div ^-2 = ^-6$$

When you divide a negative integer by a positive integer, the quotient is negative.

$$^-12 \div 2 = ^-6$$

Find the product.

1. $^-8 \times 7$ $\underline{^-56}$
2. $10 \times ^-5$ $\underline{^-50}$
3. $^-11 \times ^-2$ $\underline{22}$
4. $9 \times ^-3$ $\underline{^-27}$

5. $^-100 \times 2$ $\underline{^-200}$
6. 6×9 $\underline{54}$
7. $^-80 \times ^-10$ $\underline{800}$
8. $5 \times ^-25$ $\underline{^-125}$

9. $^-20 \times 7$ $\underline{^-140}$
10. 63×9 $\underline{567}$
11. $^-8 \times ^-17$ $\underline{136}$
12. $11 \times ^-25$ $\underline{^-275}$

Find the quotient.

13. $^-32 \div 8$ $\underline{^-4}$
14. $18 \div ^-3$ $\underline{^-6}$
15. $^-24 \div ^-6$ $\underline{4}$
16. $48 \div ^-8$ $\underline{^-6}$

17. $^-100 \div 25$ $\underline{^-4}$
18. $63 \div 9$ $\underline{7}$
19. $^-60 \div ^-3$ $\underline{20}$
20. $50 \div ^-25$ $\underline{^-2}$

Reteach **RW 57**

Name _____

Explore Operations with Rational Numbers

To add, subtract, multiply, or divide rational numbers, use the same rules you use with integers.

Add Integers	Add Rational Numbers	Multiply Integers	Multiply Rational Numbers
$3 + {}^-8 = {}^-5$	$0.4 + {}^-0.6 = {}^-0.2$	$6 \times 4 = 24$	$0.12 \times 0.2 = 0.024$
${}^-6 + {}^-4 = 10$	${}^{-1}\!/_5 + {}^{-2}\!/_5 = {}^{-3}\!/_5$	${}^-7 \times 3 = {}^-21$	${}^{-1}\!/_4 \times {}^1\!/_3 = {}^{-1}\!/_{12}$
${}^-4 + 9 = 5$	${}^-0.2 + 0.7 = 0.5$	${}^-5 \times {}^-8 = 40$	${}^-0.5 \times {}^-0.3 = 0.15$
Subtract Integers	**Subtract Rational Numbers**	**Divide Integers**	**Divide Rational Numbers**
${}^-4 - 7 = {}^-11$	${}^-0.3 - 0.8 = {}^-1.1$	$16 \div {}^-2 = {}^-8$	$0.12 \div {}^-0.3 = {}^-0.4$
${}^-8 + {}^-5 = {}^-3$	${}^{-7}\!/_8 - {}^{-2}\!/_8 = {}^{-5}\!/_8$	${}^-27 \div 3 = {}^-9$	${}^{-3}\!/_4 \div {}^1\!/_4 = {}^-3$
$11 - {}^-6 = 17$	$1.1 - {}^-0.8 = 1.9$	${}^-54 \div {}^-9 = 6$	${}^-1.5 \div {}^-0.3 = 5$

Find the sum or difference. Estimate to check.

1. ${}^-12.6 + 7.5$

2. $1\tfrac{1}{3} + {}^-5\tfrac{1}{2}$

3. ${}^-3.7 - {}^-2.4$

4. $\tfrac{9}{10} - {}^-1\tfrac{1}{5}$

_____ _____ _____ _____

5. ${}^-20.5 \div 11.8$

6. $6.7 - 9.3$

7. ${}^-5\tfrac{1}{2} + {}^-1\tfrac{1}{4}$

8. $4\tfrac{4}{5} - {}^-2\tfrac{2}{3}$

_____ _____ _____ _____

Find the product or quotient. Estimate to check.

9. ${}^-3.2 \div 0.4$

10. $1\tfrac{3}{8} \times {}^-3$

11. ${}^{-2}\!/_9 \times {}^{-6}\!/_8$

12. $4\tfrac{1}{2} \div {}^{-3}\!/_4$

_____ _____ _____ _____

13. ${}^-10.5 \div 5$

14. $0.7 \times {}^-3$

15. ${}^-6.4 \times {}^-1.5$

16. ${}^-3\tfrac{5}{6} \div 2\tfrac{2}{3}$

_____ _____ _____ _____

Name _____

LESSON 13.1

Write Expressions

Writing *numerical expressions* and *algebraic expressions* requires translating words into numbers and symbols. You can do this by looking for key words.

Addition	Subtraction	Multiplication	Division
sum	difference	product	quotient
increase	decrease	factors	equally shared
more than	less than	times	divided by
plus	minus	multiplied by	

- Write a numerical expression for "nine increased by seven squared."

  nine increased by seven squared
  $\quad\downarrow\qquad\quad\downarrow\qquad\qquad\downarrow$
  $\quad 9\qquad\quad +\qquad\qquad 7^2$

  A numerical expression contains only numbers.

- Write an algebraic expression for "nine increased by a number, n."

  nine increased by a number, n
  $\quad\downarrow\qquad\quad\downarrow\qquad\qquad\downarrow$
  $\quad 9\qquad\quad +\qquad\qquad n$

  A variable is a letter or symbol that represents an unknown number.

  An algebraic expression contains a variable.

Identify each as a *numerical expression* or *algebraic expression*. Explain your answer.

1. $12 + 8$

   __numerical expression;__

   __no variable__

2. $m - 6$

   __algebraic expression;__

   __has a variable__

3. $x \div 9$

   __algebraic expression;__

   __has a variable__

Write an algebraic expression for the word expression. Remember to look for key words.

4. thirty more than a number, n

   __$n + 30$__

5. t increased by nineteen

   __$t + 19$__

6. the product of y and twenty

   __$y \times 20$__

7. sixty divided by a number, m

   __$60 \div m$__

Reteach RW59

Name _____

LESSON 13.2

Evaluate Expressions

Evaluating expressions is like substituting players in sports. In place of a variable, you must substitute a numerical value.

Evaluate $12 \div 4 + k$, for $k = 16$.

Order of Operations
1. Operate inside parentheses.
2. Clear exponents.
3. Multiply and divide from left to right.
4. Add and subtract from left to right.

Step 1
To evaluate $12 \div 4 + k$, replace the variable with the numerical value.

$12 \div 4 + k$
$\updownarrow$
$12 \div 4 + 16$

Step 2
Use the order of operations to simplify.

$12 \div 4 + 16$ No parentheses / No exponents

$12 \div 4 + 16$ Divide: $12 \div 4 = 3$

$3 + 16$ Add: $3 + 16 = 19$

19

So, the answer is 19.

Use the order of operations to tell which operation you would perform first, second, and third. Find the value.

1. $4 \times 6 + 7 \times 2$
      multiply, multiply,
      add; 38

2. $30 \div 3 - 2 \times 4$
      divide, multiply,
      subtract; 2

3. $12 + 6 \div 2 \times 3$
      divide, multiply,
      add; 21

4. $^{-}3 + 12 \times 6 - 5$
      multiply, add,
      subtract; 64

5. $^{-}6 - 5 + 12 - 3$
      subtract, add,
      subtract; $^{-}2$

6. $^{-}2 \times {}^{-}4 \div 2 \times 3$
      multiply, divide,
      multiply; 12

Evaluate the expression for the given value of the variable.

7. $x - 13$, for $x = 20$
      7

8. $y - 30$, for $y = 10$
      $^{-}20$

9. $26 + m - 14$, for $m = 7$
      19

10. $18 + a$, for $a = {}^{-}39$
       $^{-}21$

11. $4 + k - 16$, for $k = {}^{-}5$
       $^{-}17$

12. $j \div 6 + 8$, for $j = 42$
       15

13. $5 \times (p - 15)$, for $p = {}^{-}21$
       $^{-}180$

14. $a^2 - {}^{-}3$, for $a = {}^{-}2$
       7

15. $n^2 + 8 \times 2$, for $n = 3$
       25

RW60   Reteach

LESSON 13.4

Name _____

Expressions with Squares and Square Roots

You already know pairs of operations that are opposites.
Addition and subtraction are opposites.
Addition and subtraction undo each other.
$25 + 10 = 35$, and $35 - 10 = 25$

Multiplication and division are opposites.
Multiplication and division undo each other.
$7 \times 9 = 63$, and $63 \div 9 = 7$

In the same way, squaring a number and finding the square root are also opposites.

$5^2 = 5 \times 5 = 25$ When you multiply 5 by itself, you are squaring it.
$\sqrt{25} = 5$ When you find the square root of 25, you are finding the number that can be multiplied by itself to give a product of 25.

Keep these relationships in mind when you evaluate expressions with squares and square roots.

Evaluate	$23 + \sqrt{16} \cdot (3 + 2^2)$	
	$23 + \sqrt{16} \cdot (3 + 4)$	Work inside parentheses first.
	$23 + \sqrt{16} \cdot (7)$	Continue within the parentheses.
	$23 + 4 \cdot (7)$	Evaluate $\sqrt{16}$.
	$23 + 28$	Multiply.
	51	Add.

Evaluate the expression.

1. $6 + 9^2 - 2$

   _____

2. $24 + (\sqrt{25} + 4^2)$

   _____

3. $\sqrt{36} + 9^2 - 2 \cdot \sqrt{4}$

   _____

4. $^-2 \cdot (3^2 - \sqrt{9})$

   _____

5. $\sqrt{81} - 6 + 3 \cdot (4^2)$

   _____

6. $26 + 4 \cdot \sqrt{49} - 6^2$

   _____

7. $^-5 + 4 \cdot \sqrt{9}$

   _____

8. $\sqrt{100} - 8^2 + 2 \cdot (4^2)$

   _____

9. $3^2 - \sqrt{81} + 5 \cdot (2^2)$

   _____

10. $\sqrt{64} + 9^2 - 3 \cdot (5^2)$

    _____

Reteach   **RW61**

Name _____

LESSON 14.1

Connect Words and Equations

You can write an algebraic equation for a word sentence by reading the sentence as though it were an equation.

Example 1

The price of a ticket plus $3.95 for snacks equals $10.20.

- Read the sentence aloud, listening for where you will place the variable.
- The only unknown part of this equation is the price of a ticket. Let p represent the price of a ticket.

Read as	The price	plus $3.95	equals	$10.20
Then	p	+ 3.95	=	10.20

So, the equation is $p + 3.95 = 10.20$.

Example 2

Nine-year-old Sara is one half the age of her sister, Carmen.

- In this example, you must make two decisions: where to place the variable and where to place the equal sign.
- The only unknown is Carmen's age, so let C represent her age.
- As you read the sentence aloud, think about which word or words could be replaced by the word *equals*. The word *is* represents the equal sign.

Nine-year-old Sara is one half the age of her sister, Carmen.
$$9 \quad = \quad \frac{1}{2} \quad \times \quad C$$

So, the equation is $9 = \frac{1}{2} \times C$ or $9 = \frac{C}{2}$.

Write an equation for the word sentence. Choice of variable may vary.

1. 8 more than a number is 23.
   $n + 8 = 23$

2. ⁻2 times a number is ⁻40.
   $^-2n = ^-40$

3. One third of a number is 16.
   $\frac{1}{3}(n) = 16$ or $\frac{n}{3} = 16$

4. A number divided by 5 is 7.
   $n \div 5 = 7$ or $\frac{n}{5} = 7$

5. 12 less than a number is 19.
   $n - 12 = 19$

6. 8 times a number is 56.
   $8n = 56$

RW62 Reteach

Name _____

LESSON 14.3

Solve Addition Equations

An equation tells you that two quantities are equal. You can think of an equation as a balance scale where the values on each side of the scale balance each other.

You can write an algebraic equation for a word sentence by reading the sentence as though it were an equation.

$x + 19 = 42$

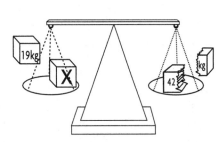

When x is alone on one side of the scale, and the scale balances, the value on the other side is the solution of the equation. So, you want to get x alone on one side of the scale.

$$\begin{aligned} x + 19 &= 42 \\ -19 &-19 \\ x &= 23 \end{aligned}$$

So, the solution to the equation $x + 19 = 42$ is $x = 23$. You can check the solution by replacing the variable in the equation with the value of x.

$$\begin{aligned} x + 19 &= 42 \\ 23 + 19 &= 42 \\ 42 &= 42 \end{aligned}$$

Solve and check.

1. $d + 17 = 22$

   $d = 5$

2. $w + 6 = 19$

   $w = 13$

3. $18 + y = 32$

   $y = 14$

4. $46 = t + 27$

   $t = 19$

5. $x + 3.6 = 10.1$

   $x = 6.5$

6. $17.2 = a + 9.7$

   $a = 7.5$

7. $p + 63 = 91$

   $p = 28$

8. $8\frac{1}{2} + k = 15$

   $k = 6\frac{1}{2}$

9. $18 = h + 6\frac{1}{3}$

   $h = 11\frac{2}{3}$

10. $5.3 + m = 10.0$

    $m = 4.7$

11. $75.3 = 32.4 + b$

    $b = 42.9$

12. $9\frac{1}{4} = a + 3\frac{1}{3}$

    $a = 5\frac{11}{12}$

Reteach RW63

Name _____

LESSON 14.4

Solve Subtraction Equations

Addition and subtraction are *inverse operations* because what one operation does, the other one can undo.

$8 + 3 = 11$

$11 - 3 = 8$ ← Subtraction undoes addition.

You can use the idea of inverse operations to solve subtraction equations.

Solve $x - 3 = 9$.

Read the equation as "a number x minus three equals nine." Since 3 is subtracted from x, *add* 3 to each side of the equation to solve for x.

$$\begin{aligned} x - 3 &= 9 \\ +3 &+3 \\ \hline x &= 12 \end{aligned}$$

Solve the equation. Show your steps.

1. $x - 5 = 16$
   $+5 +5$
   $x = 21$

2. $19 = b - 9$
   $+9 +9$
   $28 = b$

3. $y - 38 = 6$
   $+38 +38$
   $y = 44$

4. $7 = a - 15$
   $+15 +15$
   $22 = a$

5. $t - 16 = 24$
   $+16 +16$
   $t = 40$

6. $52 = p - 42$
   $+42 +42$
   $94 = p$

7. $y - 2 = 19$
   $+2 +2$
   $y = 21$

8. $23 = w - 26$
   $+26 +26$
   $49 = w$

9. $k - 36 = 21$
   $+36 +36$
   $k = 57$

10. $d - 2.5 = 6$
    $+2.5 +2.5$
    $d = 8.5$

11. $h - 3.7 = 2.5$
    $+3.7 +3.7$
    $h = 6.2$

12. $f - 11.9 = 7.3$
    $+11.9 +11.9$
    $f = 19.2$

13. $r - 3.9 = 7.8$
    $+3.9 +3.9$
    $r = 11.7$

14. $13.5 = z - 6.5$
    $+6.5 +6.5$
    $20 = z$

15. $m - 17.8 = 11.3$
    $+17.8 +17.8$
    $m = 29.1$

16. $e - \frac{1}{4} = \frac{3}{4}$
    $+\frac{1}{4} +\frac{1}{4}$
    $e = 1$

17. $c - \frac{1}{3} = 2\frac{1}{3}$
    $+\frac{1}{3} +\frac{1}{3}$
    $c = 2\frac{2}{3}$

18. $3\frac{1}{2} = k - 1\frac{1}{4}$
    $+1\frac{1}{4} +1\frac{1}{4}$
    $4\frac{3}{4} = k$

RW64 Reteach

Name _____

LESSON 15.2

Solve Multiplication and Division Equations

The expression $6b$ means $6 \times b$.
So, if $b = 8$, $6b$ does *not* mean 68. It means 6×8.
The expression $\frac{x}{5}$ means $x \div 5$.

Multiplication and division are inverse operations.
- Use division to solve a multiplication equation.
- Use multiplication to solve a division equation.

To solve $6p = 18$,
divide both sides by 6.

$$6p = 18$$
$$\frac{6p}{6} = \frac{18}{6}$$
$$p = 3$$

Always check your solution.

$$6p = 18$$
$$6 \times 3 = 18 \quad \text{Replace } p \text{ with 3.}$$
$$18 = 18 \checkmark \text{ The solution checks.}$$

To solve $\frac{y}{5} = 7$,
multiply both sides by 5.

$$\frac{y}{5} = 7$$
$$5 \times \frac{y}{5} = 7 \times 5$$
$$y = 35$$

Always check your solution.

$$\frac{y}{5} = 7$$
$$\frac{35}{5} = 7 \quad \text{Replace } y \text{ with 35.}$$
$$7 = 7 \checkmark \text{ The solution checks.}$$

Write the expression without using a multiplication symbol.

1. $8 \times b$ ___8b___
2. $m \times 4$ ___4m___
3. $15 \times x$ ___15x___
4. $a \times 9$ ___9a___

Use inverse operations to solve. Check your solution.

5. $4x = 16$

   ___x = 4___

6. $7x = 14$

   ___x = 2___

7. $4s = 12$

   ___s = 3___

8. $3y = 9$

   ___y = 3___

9. $\frac{x}{4} = 2$

   ___x = 8___

10. $\frac{y}{3} = 4$

    ___y = 12___

11. $\frac{s}{2} = 1$

    ___s = 2___

12. $\frac{x}{5} = 6$

    ___x = 30___

13. $32 = 4n$

    ___n = 8___

14. $\frac{a}{5} = 3$

    ___a = 15___

15. $8 = \frac{n}{8}$

    ___n = 64___

16. $6m = 66$

    ___m = 11___

Reteach RW65

Name _____

LESSON 15.3

Use Formulas

If the temperature is 30°, is it hot or is it cold? The answer depends on whether the temperature is measured in degrees Celsius (°C) or degrees Fahrenheit (°F).

- 30°F is cold enough for a major snowstorm. Water freezes at 32°F.
- 30°C is a typical temperature on a hot summer afternoon. At 30°C, you might feel like going swimming.

To convert from °C to °F, use the formula below.
$$F = (\tfrac{9}{5} \times C) + 32$$
To convert from °F to °C, use the formula below.
$$C = \tfrac{5}{9} \times (F - 32)$$

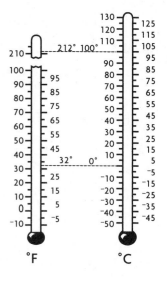

Complete.

1. Convert 30°C to degrees Fahrenheit.

   $F = (\tfrac{9}{5} \times C) + 32$

   $= (\tfrac{9}{5} \times \underline{\ 30\ }) + 32$

   $= \underline{\ 54\ } + 32$

   $= \underline{\ 86\ }$ °F

2. Convert 59°F to degrees Celsius.

   $C = \tfrac{5}{9} \times (F - 32)$

   $= \tfrac{5}{9} \times (\underline{\ 59\ } - 32)$

   $= \tfrac{5}{9} \times \underline{\ 27\ }$

   $= \underline{\ 15\ }$ °C

Convert the temperature to degrees Fahrenheit. Round the answer to the nearest degree.

3. 15°C __59°F__   4. 80°C __176°F__   5. 100°C __212°F__   6. 42°C __108°F__

7. 25°C __77°F__   8. 32°C __90°F__   9. 71°C __160°F__   10. 19°C __66°F__

Convert the temperature to degrees Celsius. Round the answer to the nearest degree.

11. 41°F __5°C__   12. 50°F __10°C__   13. 200°F __93°C__   14. 89°F __32°C__

15. 212°F __100°C__ 16. 38°F __3°C__   17. 53°F __12°C__   18. 69°F __21°C__

Name _____

LESSON 15.5

Problem Solving Strategy: Work Backward

Jayne is going to hike the Appalachian Trail. To prepare, she bought 10 lb of hiker's trail mix, a stuff sack for $14.60, and a water filter for $59.99. She spent a total of $91.49. How much did the trail mix cost per pound?

Step 1 Write the information you know.

total amount spent ($91.49)

cost of two items purchased ($14.60 and $59.99)

want to find the cost of the trail mix and then the price per pound

Step 2 Plan a strategy to solve. Use the strategy *work backward*.
- Write an equation using the information in the problem.

| cost of trail mix ? | + | stuff sack ($14.60) | + | water filter ($59.99) | = | total spent ($91.49) |

- Decide how you can find the missing information. Write the information in reverse order with inverse operations.

total spent − water filter − stuff sack = cost of trail mix
  $91.49 − $59.99 − $14.60 = ?

cost of trail mix ÷ 10 = price per pound of trail mix

Step 3 Solve. Carry out the strategy.

$91.49 − $59.99 − $14.60 = $16.90

$16.90 ÷ 10 = $1.69

The trail mix cost $1.69 per lb.

Use the strategy *work backward* to help you solve.

1. Marco bought 5 yd of fabric to make kites. He also paid $24.89 for rods and $17.00 for string. Marco spent $98.49 in all. What was the cost of the fabric per yard?

   _____ $11.32 _____

2. The print store charges $0.55 for each transparency made plus a $25.00 service charge. Sarah spent $135 on her project. How many transparencies did she make?

   _____ 200 transparencies _____

3. If you add 7 to Jason's age, then multiply by 5, the result is 105. How old is Jason?

   _____ 14 years old _____

4. If you subtract 10 from Sylvia's age, then multiply by 4, the result is 216. How old is Sylvia?

   _____ 64 years old _____

Reteach RW67

Name _____

Points, Lines, and Planes

Here is a drawing of the front wall in a typical classroom. It shows a chalkboard, flag, bulletin board, clock, and poster.

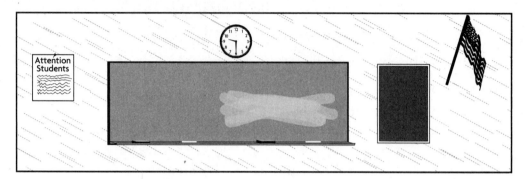

Here are some geometric figures. The arrowhead means the figure goes on forever in that direction.

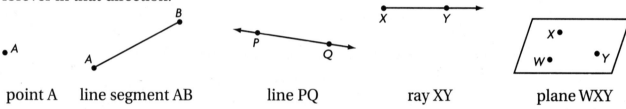

point A   line segment AB   line PQ   ray XY   plane WXY

Name the geometric figure suggested by each item on the classroom wall.

1. the flag pole _____line segment_____

2. where the hands of the clock meet _____point_____

3. the "Attention Students" poster _____plane_____

4. the chalkboard _____plane_____

5. where the flag pole is fastened to the wall _____point_____

6. a piece of chalk _____line segment_____

7. a corner of the bulletin board _____point_____

8. the left edge of the chalkboard _____line segment_____

9. a hand on the clock _____ray or line segment_____

10. the chalk ledge _____plane_____

RW68   Reteach

Name _____

Angle Relationships

Find the measure of angles ∠2, ∠3, and ∠4.

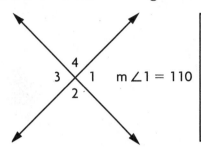

Definitions
Angles are formed when two lines intersect. The opposite angles, called *vertical angles,* have the same measure.
Supplementary angles are two angles that form a 180° angle.

m∠1 = 110

Step 1 Use the definition of *vertical angles.* Since ∠3 is opposite ∠1, then m∠3 = m∠1. So ∠3 = 110°.

Step 2 Use the definition of *supplementary angles* to find the measure of ∠4.

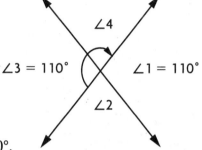

Angles 3 and 4 are supplementary angles. Their sum equals 180°. To find the measure of ∠4, subtract the measure of ∠3 from 180°.

180° − 110° = 70°   So, m∠4 = 70°.

Step 3 Use the definition of *vertical angles* to find the measure of ∠2.

Since ∠4 is opposite ∠2, then m∠4 = m∠2. So m∠2 = 70°.

The angle measures are m∠2 = 70°, m∠3 = 110°, and m∠4 = 70°.

Find the unknown measures in the following figures.

1.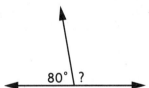

2.

3.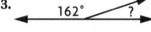

4.

5.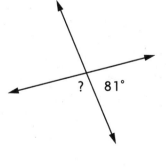

6.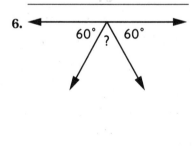

Reteach **RW69**

Name _____

LESSON 16.4

Classify Lines

Learning math vocabulary is easier if you connect it to everyday language.

Explain what is meant by each underlined word in the following statements.

- There is a light at the <u>intersection</u>.

  The light is located where two or more streets meet.

- Be sure the edges of the picture frame are <u>perpendicular</u>.

  The edges meet at a 90° angle.

- Some people find <u>parallel</u> parking difficult.

  Parking so that the car aligns with the curb can be difficult.

Here are the math definitions for these words.

Intersecting lines are lines that cross at exactly one point.

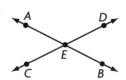

Lines CD and AB intersect at point E.

Perpendicular lines are lines that intersect to form 90° angles, or right angles.

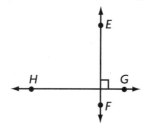

Lines EF and GH are perpendicular.

Parallel lines are lines that are always the same distance apart.

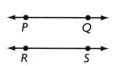

$\overleftrightarrow{PQ} \parallel \overleftrightarrow{RS}$

Lines PQ and RS are parallel.

For Exercises 1–5, use the figure at the right.

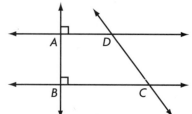

1. Name a line that is parallel to $\overleftrightarrow{AD}$. ____ BC ____

2. Name a line that is perpendicular to $\overleftrightarrow{BC}$. ____ AB ____

3. Name two lines that intersect $\overleftrightarrow{AD}$. ____ AB and DC ____

4. Name a line that intersects $\overleftrightarrow{BC}$ but is not perpendicular to $\overleftrightarrow{BC}$. ____ DC ____

5. $\overleftrightarrow{AB}$ and $\overleftrightarrow{BC}$ intersect at what point? ____ point B ____

RW70 Reteach

Name _____

LESSON 17.1

Triangles

There are several facts that you will need to use in your study of triangles:

- A right angle has a measure of 90°.
- A straight angle has a measure of 180°.
- The sum of the measures of the angles of a triangle is 180°.

Draw any right triangle.

Fold each of the acute angles onto the right angle. The sum of the two acute angles should match the right angle.

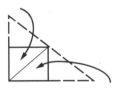

Since the right angle has a measure of 90°, the sum of the two acute angles also is 90°. The sum of the measures of the three angles is 180°.

Triangles can be classified according to the measures of their angles.

- In an **acute triangle,** all angles will be less than 90°.
- In an **obtuse triangle,** there will be one angle greater than 90°.
- In a **right triangle,** there will be one 90° angle.

Find the measure of the angle and classify the triangle.

1.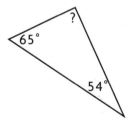

   _____

2.

   _____

3.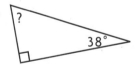

   _____

4.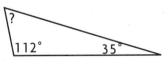

   _____

5.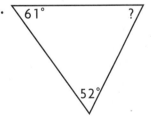

   _____

6.

   _____

Reteach **RW71**

Name _____

Problem Solving Strategy: Find a Pattern

Juan has a piece of land that he wants to separate into sections. He wants to be able to keep the animals he raises apart from one another. The sections can be different sizes to hold different-size animals.

What is the greatest number of sections Juan can make if he builds only fences that are straight and he uses 6 fences?

Step 1 Think about what you know and what you are asked to find.
- You know that the sections do not have to be the same size.
- You know that Juan will use only straight fences.
- You are asked to find the greatest number of sections he can make from 6 fences.

Step 2 Plan a strategy to solve.
- Use the strategy *find a pattern*.
- Look for a pattern as the number of fences increases, starting with 1 fence.

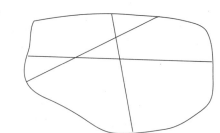

Step 3 Solve.
- Carry out the strategy.
Make a table. Use the data in the table to look for a pattern.

Number of Fences	Number of Sections	Increase
1	2	
2	4	2
3	7	3
4	11	4

The pattern appears to be that for each fence Juan adds, he increases the number of sections by one more than the previous increase.

So, for 5 fences, Juan can add 5 sections to the 11 he has, for a total of 16. For 6 fences, he can add 6 sections to 16, for a total of 22.

Use the strategy *find a pattern* to help you solve.

1. The town library has a front stairway with 5 steps. The first step is made from 6 stone blocks. Each step uses 3 more stone blocks than the previous step. How many blocks are used in the stairway?

   _____ 60 blocks _____

2. A display of cans in a store window has 9 cans on the bottom row. The number of cans decreases by 1 with each row. The top row is 1 can. How many cans are in the entire display?

   _____ 45 cans _____

RW72 Reteach

Name _____

LESSON 17.3

Quadrilaterals

You have seen that quadrilaterals can be related to one another by use of a Venn diagram.

The relationship between quadrilaterals can also be shown by another type of diagram.

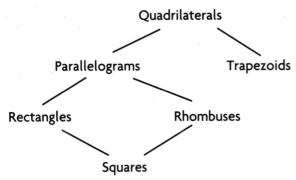

The diagram shows the following:
- All the figures are quadrilaterals.
- All rectangles are parallelograms.
- All rhombuses are parallelograms.
- All squares are rectangles.
- All squares are rhombuses.
- A trapezoid is neither a parallelogram nor a rectangle, rhombus, or square.

Use the diagram at the top of the page. Complete each sentence using *must be*, *can be*, or *cannot be*.

1. a rhombus ____can be____ a square.

2. a rectangle ____must be____ a parallelogram.

3. a trapezoid ____cannot be____ a rhombus.

4. a parallelogram ____can be____ a rhombus.

5. a rectangle ____can be____ a square.

6. a square ____must be____ a rhombus.

Reteach RW73

Name _____

LESSON 17.4

Draw Two-Dimensional Figures

In order to draw a two-dimensional figure, use the following steps:
- Recall the properties of the figure you are going to draw.
- Decide on the order in which you will use those properties.

Remember, when you draw a figure on dot paper, you may extend the sides past the point where they intersect.

Draw a parallelogram with no right angles.

Step 1 Recall the properties of the figure.
- A parallelogram is a quadrilateral.
- Opposite sides are congruent.
- Opposite sides are parallel.

Step 2 Decide on the order in which you will use these properties.
- On dot paper, draw two parallel line segments. It is not important for them to be the same length.
- Now, draw the second pair of sides, making them parallel.

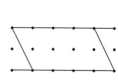

Draw the figure. Use square dot paper or isometric dot paper. Check students' drawings.

1. an acute scalene triangle

2. a pentagon with three congruent sides

3. a quadrilateral with four congruent angles and four congruent sides

4. a hexagon with all sides congruent

5. a triangle with one obtuse angle and two congruent sides

6. a quadrilateral with two pairs of congruent sides and a right angle

RW74 Reteach

Name _____

LESSON 17.5

Circles

To better understand a circle, draw one. You can construct a circle using just a pencil and a ruler.

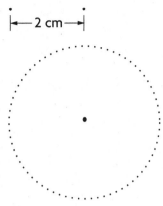

- Draw a dot and measure a length of 2 cm from the dot. Mark another dot 2 cm from the first one.
- Continue to measure 2 cm lengths from the first dot you made. Mark each location with another dot. You'll need to make at least 20 marks.

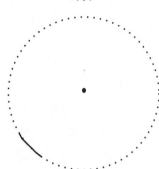

- As you fill in the spaces, you will notice that the dots are becoming closer and closer together. If you continued to fill in the spaces between the dots with more dots, the line segments connecting them would get so short that eventually you could not see them. When the space between the marks was completely filled, you would have a circle.

- A **circle** is all the points in a plane that are the same distance form one point, the center.
- The lengths that you measured are all **radii** of the circle.
- The distance from one side of the circle to the other, through the center, is the **diameter** of the circle.
- The distance between any two dots along the circle is an **arc.**

Use the circle you drew to answer 1–3.

1. How many points do there seem to be around the circle?

   _____ an infinite number _____

2. How many radii do there seem to be around the circle?

   _____ an infinite number _____

3. How many diameters do there seem to be around the circle?

   _____ an infinite number _____

Name _____

Types of Solid Figures

A **polyhedron** is a solid figure with flat faces that are polygons. A polyhedron is named by the shape of its base.

A **prism** is a polyhedron with two congruent and parallel bases. In a prism the lateral faces between the bases are rectangles.

Name this figure.
Think: All the faces are flat. All the faces are polygons.
There are two congruent and parallel bases.
The base is a triangle, and the lateral faces are rectangles.
So, the figure is a triangular prism.

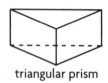
triangular prism

A **pyramid** is a polyhedron that has one base, and the lateral faces are triangles that meet at a vertex.

Name this figure.
Think: All the faces are flat. All the faces are polygons.
The lateral faces are triangles that meet at a vertex.
The base is a square.
So, the figure is a square pyramid.

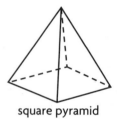
square pyramid

Name each solid figure.

1.
   triangular pyramid

2.
   rectangular prism

3.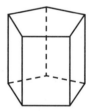
   pentagonal prism

4.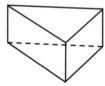
   triangular prism

5.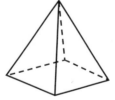
   square pyramid

6.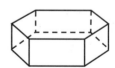
   hexagonal prism

RW76 Reteach

Name _____

LESSON 18.2

Different Views of Solid Figures

Solid figures can look different when you view them from the top, front, and side.

Look at the rectangular prism at the right.

If you look at the prism only from the top, you would see a rectangle.

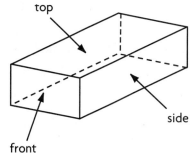

If you look at the prism only from the front, you would see a rectangle.

If you look at the prism only from the side, you would see a rectangle.

Name the figure you would see from each view of the solid figure.

1. Top: _____triangle_____

   Front: _____triangle_____

   Side: _____triangle_____

2. Top: _____square_____

   Front: _____square_____

   Side: _____square_____

3. Top: _____pentagon_____

   Front: _____triangle_____

   Side: _____triangle_____

Reteach RW77

Name _____

Problem Solving Strategy: Solve a Simpler Problem

Rhea is building models of prisms, using balls of clay for vertices and straws for edges. How many balls of clay and how many straws will she need to make a prism whose base has 8 sides?

Sometimes it helps you solve a more difficult problem by solving a similar, but simpler, problem first.

You can *solve a simpler problem* by thinking about the numbers of vertices and edges on prisms whose bases have 3, 4, and 5 sides.

Number of sides on base	3	4	5
Number of vertices	3 + 3, or 6	4 + 4, or 8	5 + 5, or 10
Number of edges	3 + 3 + 3, or 9	4 + 4 + 4, or 12	5 + 5 + 5, or 15

From the pattern in the table, you can see that for a prism whose base has 8 sides, the number of vertices is 8 + 8, or 16, and the number of edges is 8 + 8 + 8, or 24.

So, Rhea will need 16 balls of clay and 24 straws.

Solve by first solving a simpler problem.

1. Suppose Rhea is building a model of a prism whose base has 16 sides. How many balls of clay and how many straws will she need? _____

2. Greg wants to make a model of a pyramid whose base has 5 sides. He will use balls of clay for the vertices and toothpicks for the edges. How many toothpicks will he need? _____

3. Terri is making edible prisms. She uses cheese cubes for vertices and carrot sticks for edges. How many cheese cubes and carrot sticks will she need to make a prism whose base has 14 sides? _____

RW78 Reteach

LESSON 19.1

Name _____

Construct Congruent Segments and Angles

Here are two constructions you can do with a compass and a straightedge.

1. Construct a line segment congruent to $\overline{AB}$.

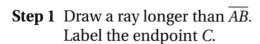

Step 1 Draw a ray longer than $\overline{AB}$.
Label the endpoint C.

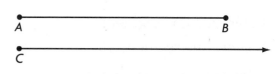

Step 2 Measure $\overline{AB}$ using your compass.
Put the point on A, and open the
compass so the pencil is on B.

Step 3 Use the same opening as in Step 2.
Put the compass point on C.
Draw an arc intersecting the ray.
Label the intersection point D. $\overline{AB} \cong \overline{CD}$.

2. Construct an angle congruent to $\angle ABC$.

Step 1 With the compass point on B,
draw an arc through $\angle ABC$.

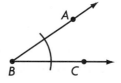

Step 2 Draw $\overrightarrow{DE}$. Use the same opening as in Step 1. With the
compass point on D, draw an arc just like the one you
drew in Step 1.

Step 3 Use the compass to measure the arc in $\angle ABC$.

Step 4 Using the same opening, locate point F on the arc.

Step 5 Draw $\overrightarrow{DF}$. $\angle ABC \cong \angle FDE$.

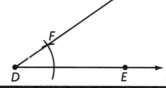

---

1. Use a compass and a straightedge to construct a line segment congruent to $\overline{PQ}$. Check students' drawings.

2. Use a compass and a straightedge to construct an angle congruent to $\angle RST$. Check students' drawings.

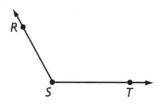

Reteach **RW79**

Name _____

LESSON 19.2

Bisect Line Segments and Angles

1. Fill in the boxes in each diagram with the number of the step that was shown.

To bisect a line segment:
1. Draw line segment AB.
2. Place the compass point on point A.
3. Open the compass to a length greater than half the length of $\overline{AB}$, and draw an arc through $\overline{AB}$.
4. Keep the same compass opening and place the compass point on point B.
5. Draw an arc through $\overline{AB}$.
6. Label the points C and D where the arcs intersect.
7. Draw line CD.

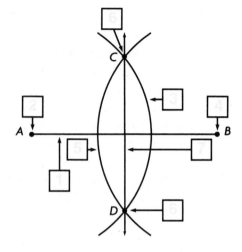

To bisect an angle:
1. Place the compass point on point B.
2. Draw an arc through $\overrightarrow{BC}$ and $\overrightarrow{BA}$.
3. Label the points of intersection D and E.
4. Place the compass point on point D and draw an arc.
5. Using the same opening, place the compass point on point E and draw an arc.
6. Label the point of intersection of the two arcs point F.
7. Draw $\overrightarrow{BF}$ as the bisector of $\angle ABC$.

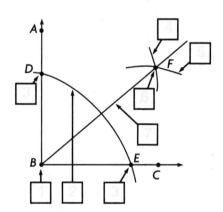

---

2. Construct the bisector of line segment XY. Check students' drawings.

3. Construct the bisector of $\angle JKL$. Check students' drawings.

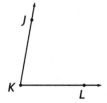

Name _____

LESSON 19.4

Similar and Congruent Figures

Looking in the mirror, you see an exact image of yourself. You and your image are congruent shapes.

Trick mirrors can make people appear out of proportion. Look at the image of each person in the mirrors below.

The person and the image are similar. They have the same shape.

The person and the image are not similar. They have the same height but different widths.

The person and the image are not similar. They have the same width but different heights.

Tell whether the figures in each pair are *similar, congruent, both,* or *neither*.

1.
   neither
   _____

2.
   both
   _____

3.
   similar
   _____

4.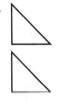
   both
   _____

5. similar
   _____

6.
   neither
   _____

7.
   neither
   _____

8.
   similar
   _____

9.
   both
   _____

Reteach RW81

Name _____   **LESSON 20.1**

Ratios and Rates

A **ratio** compares two numbers. The ratio of
shaded squares to unshaded squares is two to
four. It can be written in three ways:

2 to 4   2:4 $\frac{2 \leftarrow \text{first term}}{4 \leftarrow \text{second term}}$

Now, divide the figure in half. The amount of
shading hasn't changed. The ratio of shaded
parts to unshaded parts is now 4 to 8.

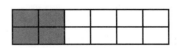

$\frac{2}{4}$ and $\frac{4}{8}$ are **equivalent ratios** because they show the same relationship between
shaded and unshaded parts. If you multiply or divide both terms of a ratio by
the same number, the result is an equivalent ratio.

$\frac{2}{4} \rightarrow \frac{2 \times 2}{4 \times 2} = \frac{4}{8}$ $\frac{2}{4}$ and $\frac{4}{8}$ are equivalent. $\frac{15}{10} \rightarrow \frac{15 \div 5}{10 \div 5} = \frac{3}{2}$ $\frac{15}{10}$ and $\frac{3}{2}$ are equivalent.

The word **rate** is used for ratios that compare quantities having
different units of measure, such as miles per gal, miles per hr, or
apples per dollar.

A price of $30 per dozen pens can be written as a rate.

$\frac{\$30}{12 \text{ pens}}$ ← This ratio is a rate, comparing dollars to pens.

A **unit rate** is a rate in which the second term is 1.

You can write the above rate as a unit rate.

first term → $\frac{\$30}{12}$ = $\frac{\$30 \div 12}{12 \div 12}$ ← Divide both terms by 12
second term →   to get 1 in the denominator.

So, at a rate of $30 for 12 pens, the unit rate is $2.50 per pen.

Write three equivalent ratios. Possible answers are given.

1. $\frac{3}{6}$ 1, 6, 9 / 2, 12, 18   2. $\frac{7}{1}$ 14, 21, 28 / 2, 3, 4   3. $\frac{4}{16}$ 1, 2, 8 / 4, 8, 32

4. Which of the following ratios are rates? _____ b, c

   a. $\frac{\$3}{\$2}$ b. $\frac{40 \text{ people}}{10 \text{ cars}}$ c. $\frac{\$20}{5 \text{ lb}}$ d. $\frac{25 \text{ mi}}{15 \text{ mi}}$

Write each ratio in fraction form. Then find the unit rate.

5. $9.00 for 6 gal   6. 300 mi per 10 gal   7. 200 mi per 4 hr

   $\frac{\$9.00}{6 \text{ gal}}$; $1.50 per gal $\frac{300 \text{ mi}}{10 \text{ gal}}$; 30 mi per gal $\frac{200 \text{ mi}}{4 \text{ hr}}$; 50 mi per hr

RW 82   Reteach

Name _____

LESSON 20.3

Problem Solving Strategy

Write an Equation

An equation with a ratio on each side is a **proportion**. For example, $\frac{1}{2} = \frac{5}{10}$ and $\frac{1}{n} = \frac{3}{9}$ are proportions.

When the proportion has a variable for one of the terms, you can solve the proportion by cross-multiplying. Study this problem.

Example: A car travels 75 mi on 3 gal of gasoline. How far can it go on 8 gallons?

- Write a proportion. You can write the ratios in several ways:

$$\frac{\text{mi}}{\text{mi}} = \frac{\text{gal}}{\text{gal}} \text{ or } \frac{\text{gal}}{\text{mi}} = \frac{\text{gal}}{\text{mi}} \text{ or } \frac{\text{mi}}{\text{gal}} = \frac{\text{mi}}{\text{gal}}$$

Compare miles to gallons.

$\frac{75 \text{ mi}}{3 \text{ gal}} = \frac{n \text{ mi}}{8 \text{ gal}}$ ← Let the variable stand for the unknown distance.

- Solve the proportion by cross-multiplying.

$\frac{75}{3} \bowtie \frac{n}{8}$

$3n = 75 \times 8$

$3n = 600$ ← Divide by 3.

$n = 200$

The car can go 200 mi on 8 gal of gasoline.

Solve the problem by writing an equation.

1. You can bicycle 5 mi in 40 min. At that rate, how long does it take to travel 18 mi?

   _____ 144 min, or 2 hr 24 min _____

2. Six cartons of bread weigh 192 lb. How much do 10 cartons weigh?

   _____ 320 lb _____

3. A car travels 360 mi on 12 gal of gasoline. How far can it go on 15 gal?

   _____ 450 mi _____

4. A plane travels 200 mi in 30 min. At that rate, how long does it take to travel 900 mi?

   _____ 135 min, or 2 hr 15 min _____

5. A straight fence requires 4 posts for every 50 ft of fence. How many posts are required for 325 ft of fence?

   _____ 26 posts _____

6. Eight candles cost $2.00. What is the cost of 36 candles?

   _____ $9.00 _____

Reteach   RW 83

Algebra: Ratios and Similar Figures

Two shapes that are similar have corresponding sides and corresponding angles. Triangles *DEF* and *ABC* are similar.

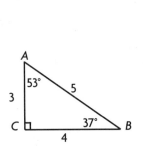
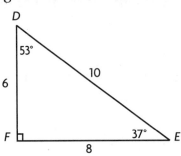

Corresponding Sides

$\overline{AB}$ corresponds to $\overline{DE}$.
$\overline{BC}$ corresponds to $\overline{EF}$.
$\overline{CA}$ corresponds to $\overline{FD}$.

The corresponding sides have the same ratio.

$\dfrac{AB}{DE} = \dfrac{5}{10} = \dfrac{1}{2}$

$\dfrac{BC}{EF} = \dfrac{4}{8} = \dfrac{1}{2}$

$\dfrac{CA}{FD} = \dfrac{3}{6} = \dfrac{1}{2}$

Corresponding Angles

$\angle A$ corresponds to $\angle D$.
$\angle B$ corresponds to $\angle E$.
$\angle C$ corresponds to $\angle F$.

The corresponding angles are equal.

$\angle A = 53°$ and $\angle D = 53°$

$\angle B = 37°$ and $\angle E = 37°$

$\angle C = 90°$ and $\angle F = 90°$

Name the corresponding sides and angles. Write the ratio of the corresponding sides in simplest form.

1.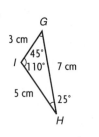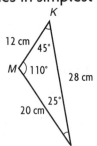

2.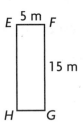

GH corresponds to *KL*; *HI* corresponds to *LM*; *IG* corresponds to *MK*; $\angle G$ corresponds to $\angle K$; $\angle H$ corresponds to $\angle L$; $\angle I$ corresponds to $\angle M$; $\dfrac{1}{4}$ or $\dfrac{4}{1}$

AB corresponds to *EF*; *BC* corresponds to *FG*; *CD* corresponds to *GH*; *DA* corresponds to *HE*; $\angle A$ corresponds to $\angle E$; $\angle B$ corresponds to $\angle F$; $\angle C$ corresponds to $\angle G$; $\angle D$ corresponds to $\angle H$; $\dfrac{4}{5}$ or $\dfrac{5}{4}$

RW84 Reteach

Name _____

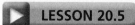

Algebra: Proportions and Similar Figures

There are objects that can't be measured by climbing them. To measure these objects, you can use indirect measurement.

Look at the drawing of a tree and its shadow. You may not be able to climb the tree, but you can measure the shadow.

You could also measure your own height and your shadow.

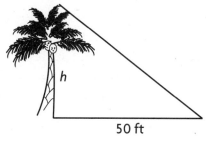

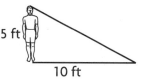

Using similar figures, you can find the height of the tree. Follow these steps.

Step 1
Write ratios using corresponding sides.

$\dfrac{\text{your shadow}}{\text{tree's shadow}} = \dfrac{10 \text{ ft}}{50 \text{ ft}}$

$\dfrac{\text{your height}}{\text{tree's height}} = \dfrac{5 \text{ ft}}{h}$

Step 2
Set the ratios equal to each other.

$\dfrac{10 \text{ ft}}{50 \text{ ft}} = \dfrac{5 \text{ ft}}{h}$

Step 3
Solve the proportion.

$\dfrac{10 \text{ ft}}{50 \text{ ft}} = \dfrac{5 \text{ ft}}{h}$

$10 \times h = 50 \times 5$

$\dfrac{10h}{10} = \dfrac{250}{10}$

$h = 25$

So, the tree is 25 ft tall.

Use the two similar right triangles to write a proportion. Then solve for h.

1.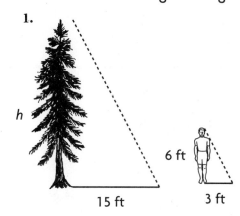

2.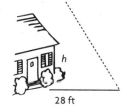

3.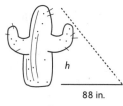

$\dfrac{h}{6} = \dfrac{15}{3}$; $h = 30$ ft $\dfrac{h}{7} = \dfrac{28}{4}$; $h = 49$ ft $\dfrac{h}{52} = \dfrac{88}{44}$; $h = 104$ in.

Reteach   **RW85**

Name _____

LESSON 20.6

Algebra: Scale Drawings

Tom's backyard is a rectangle 100 ft long and 50 ft wide. Measure the rectangles below. Which of them can be a scale drawing of Tom's backyard? What is the scale?

a.    b.    c.

Only figure **b** can be a scale drawing of Tom's backyard. Figure **a** is too narrow, and figure **c** is too square. Tom's backyard is twice as long as it is wide, and so is figure **b**.

In a scale drawing, the shape is the same as the actual object. The drawing is smaller or larger.

Since figure **b** is 1 in. wide and Tom's yard is 50 ft wide, the drawing has a scale of 1 in. = 50 ft.

Two trees in the yard are $1\frac{1}{2}$, or 1.5, in. apart on the drawing. How far apart are the actual trees? Write a proportion.

$$\frac{1}{50} = \frac{1.5}{d}$$
$$d \times 1 = 50 \times 1.5$$
$$d = 75$$

← { Compare inches to feet.
The scale is $\frac{1 \text{ in.}}{50 \text{ ft}}$.
The distance between the trees is d ft. }

The trees are 75 ft apart.

Find the unknown dimension.

1. scale: 1 in. = 2 ft
   drawing length: 5 in.

   actual length: __10__ ft

2. scale: 1 in. = 6 ft
   drawing length: 2 in.

   actual length: __12__ ft

3. scale: 1 in. = 20 ft

   drawing length: __5__ in.
   actual length: 100 ft

4. scale: 10 cm = 3 mm
   drawing length: 5 cm

   actual length: __1.5__ mm

RW 86 Reteach

Name _____

LESSON 20.7

Algebra: Maps

A map is a scale drawing of a region. You use the scale on a map to write a proportion in the same way you use the scale in a scale drawing.

How far is it from point A to point B?

The distance between the points on the map is 2 in. You know the scale, so you can write a proportion.

$\frac{in.}{mi} \rightarrow \frac{1}{5} = \frac{2}{n} \leftarrow \frac{in.}{mi}$

$1 \times n = 2 \times 5$

$n = 10$

The distance from point A to point B is 10 mi.

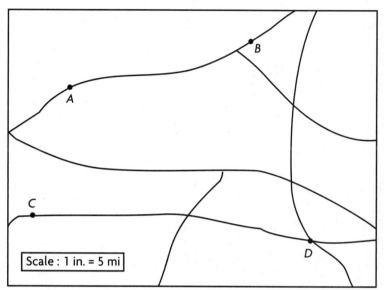

Scale : 1 in. = 5 mi

---

1. On the map above, the distance from point C to point D is 3 in. What is the actual distance from point C to point D?

   a. Complete the proportion.

   $\frac{1}{5} = \frac{\boxed{3}}{n}$

   b. Solve the proportion. $n = $ ___15___

   c. So, the distance from point C to point D is ___15 mi___.

Write and solve a proportion to find the actual distance. Use the scale 1 in. = 5 mi.

2. map distance: 1 in.

   $\frac{1}{5} = \frac{4}{n}$; 20 mi
   _____

3. map distance: $2\frac{1}{2}$ in.

   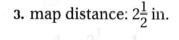
   _____

4. map distance: 6 in.

   _____

5. map distance: $7\frac{1}{2}$ in.

   $\frac{1}{5} = \frac{7\frac{1}{2}}{n}$; $37\frac{1}{2}$ mi
   _____

Reteach RW87

Name _____

LESSON 21.1

Percent

What percent of the figure is shaded?
To find out, make a fraction.

$\frac{9}{30}$ ← 9 parts are shaded.
← There are 10 × 3, or 30, equal parts in all.

Now, write the fraction as a decimal by dividing 30 into 9.

```
     0.30
30)9.00
   -9 0
     00
```

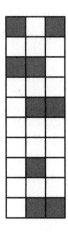

Finally, write the decimal as a percent.

0.30 = 30%   ← Move the decimal point 2 places to the right.

So, 30% of the figure is shaded.

Tell what percent of the figure is shaded.

1. 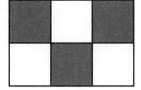   2. 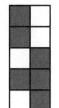   3.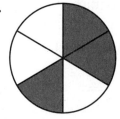

_____ 50%   _____ 60%   _____ 50%

4. 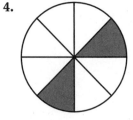   5. 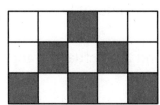   6.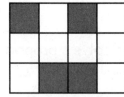

_____ 25%   _____ 40%   _____ $33\frac{1}{3}\%$

7. 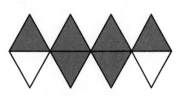   8. 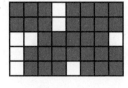   9.

_____ 75%   _____ 80%   _____ 62.5%

RW 88   Reteach

Name _____

LESSON 21.2

Percents, Decimals, and Fractions

Percent means "per 100." So, 25% means $\frac{25}{100}$.
A **percent** is the ratio of a number to 100. You can write any percent as a fraction or decimal.

- To write a percent as a fraction, put the percent over 100 and drop the % sign. Write the fraction in simplest form.

  $60\% = \frac{60}{100} = \frac{3}{5}$

- To write a percent as a decimal, move the decimal point two places to the left. Drop the percent sign. Remember, for a whole-number percent, the decimal point is to the right. So, 85% = 85.0%.

  $85\% = 0.85$

You can write decimals and fractions as percents.

- To write a decimal as a percent, move the decimal point two places to the right. Add a % sign.

  $0.255 = 25.5\%$

- To write a fraction as a percent, first write the fraction as a decimal. Then write the decimal as a percent.

  $\frac{5}{8} = 0.625$ ← Divide the numerator by the denominator.
  $= 62.5\%$

Complete the table.

	Fraction	Decimal	Percent
1.	$\frac{1}{2}$	0.5	50%
2.	$\frac{1}{5}$	0.2	20%
3.	$\frac{3}{4}$	0.75	75%
4.	$\frac{9}{10}$	0.9	90%
5.	$\frac{3}{20}$	0.15	15%
6.	$\frac{3}{8}$	0.375	37.5%

Reteach RW89

Name _____

LESSON 21.3

Estimate and Find Percent of a Number

Jill surveyed 20 of her classmates about creating a reading period at school. She found that 55% of those surveyed favored having a special time for reading during the day. How many of her 20 classmates favored a reading period, or what is 55% of 20?

Solving percent problems can be done using this formula and equivalent fractions. $\dfrac{\%}{100} = \dfrac{\text{part}}{\text{whole}}$

Step 1
Fill in the information in the formula. % = 55 and whole = 20 students

$\dfrac{55}{100} = \dfrac{\text{part}}{20}$

Step 2
Find what 100 has to be divided by to give a result of 20.

Think: $100 \div 20 = 5$.

Step 3
Divide 55 by 5 also.

$\dfrac{55 \div 5}{100 \div 5} = \dfrac{11}{20}$

So, 11 students favor a special reading time.

Use the formula to find the percent of the number.

1. 15% of 20

   $\dfrac{15}{100} = \dfrac{3}{20}$

2. 30% of 50

   $\dfrac{30}{100} = \dfrac{15}{50}$

3. 44% of 25

   $\dfrac{44}{100} = \dfrac{11}{25}$

4. 20% of 5

   $\dfrac{20}{100} = \dfrac{1}{5}$

5. 20% of 50

6. 30% of 25

7. 40% of 5

8. 60% of 20

9. 80% of 50

   _____40_____

10. 65% of 20

    _____13_____

11. 72% of 50

    _____36_____

12. 86% of 25

    _____21.5_____

13. 20% of 15

    _____3_____

14. 40% of 20

    _____8_____

15. 80% of 9

    _____7.2_____

16. 55% of 40

    _____22_____

17. 85% of 40

    _____34_____

18. 75% of 20

    _____15_____

19. 35% of 60

    _____21_____

20. 74% of 55

    _____40.7_____

21. 75% of 80

    _____60_____

22. 45% of 60

    _____27_____

23. 85% of 100

    _____85_____

24. 25% of 25

    _____6.25_____

RW90   Reteach

Name _____

LESSON 21.5

Discount and Sales Tax

Fred wants to buy a new shirt. He found one on sale at the local store. How much will Fred pay for the shirt?

Shirt Sale
Regular Price
$25.60
30% OFF

To find the sale price of the shirt, follow these steps:

Step 1: Change 30% to a decimal.
  30% = 0.30
  Remember, move the decimal point two places to the left.

Step 2: Multiply the percent, as a decimal, by the regular price.
  $25.60 × 0.30 = $7.68
  Round to the nearest cent if needed.

Step 3: Subtract the discount from the regular price.
  $25.60 − $7.68 = $17.92
  So, Fred will pay $17.92 for the shirt.

Find the amount of discount.

1. regular price: $16.00
   30% off
   $4.80

2. regular price: $27.50
   SAVE 40%
   $11.00

3. regular price: $39.00
   60% discount
   $23.40

4. regular price: $85.50
   50% off
   $42.75

Find the sale price.

5. regular price: $26.50
   Discount 10%
   $23.85

6. regular price: $47.80
   70% off
   $14.34

7. regular price: $86.00
   SAVE 25%
   $64.50

8. regular price: $184.00
   SALE 90% off
   $18.40

Reteach **RW91**

Name _____

LESSON 21.6

Simple Interest

Kirk wanted to open a savings account with the $2,000 he had earned over the summer. If the bank's simple interest rate is 5%, how much will he have in 1 year?

The money Kirk put in the bank is called **principal**. The money the bank pays Kirk is called **interest**.

Follow these steps to find the interest Kirk will have after 1 year.

Step 1: Change 5% to a decimal.

$5\% = 0.05$

Remember, move the decimal point two places to the left.

Step 2: Multiply the percent, as a decimal, by the principal.

$\$2{,}000 \times 0.05 = \$100.00$

Round to the nearest cent if needed.

Step 3: Multiply by the number of years.

$\$100.00 \times 1 = \$100.00$

So, Kirk will receive $100.00 in interest for 1 year.

Since Kirk earned $100 in interest, at the end of the year he will have $2,100. This is his principal plus the interest earned.

Find the simple interest.

	Principal	Yearly Rate	Interest for 1 Year	Interest for 2 Years
1.	$90	4%	$3.60	$7.20
2.	$225	3.8%	$8.55	$17.10
3.	$480	5%	$24.00	$48.00
4.	$750	4.8%	$36.00	$72.00
5.	$2,600	5.4%	$140.40	$280.80
6.	$3,400	7%	$238.00	$476.00
7.	$4,900	8%	$392.00	$784.00
8.	$6,800	8.6%	$584.80	$1,169.60
9.	$12,000	7.75%	$930.00	$1,860.00
10.	$15,500	9.5%	$1,472.50	$2,945.00

RW92 Reteach

Name _____

LESSON 22.1

Theoretical Probability

To find the probability, write a ratio.

$$\text{Probability: } \frac{\text{number of favorable outcomes}}{\text{number of possible equally likely outcomes}}$$

---

1. You have a spinner numbered 1 to 5. What is the probability of spinning a 2?

   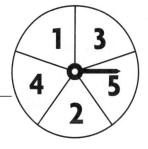

   a. How many possible outcomes are there? List them.

   _____ 5; 1, 2, 3, 4, 5 _____

   b. How many favorable outcomes are there? ____1____

   c. You can write as a fraction the probability of spinning a 2.

   $$P(2) = \frac{\text{number of favorable outcomes}}{\text{number of possible equally likely outcomes}}$$

   Write the probability as a fraction. ____$\frac{1}{5}$____

2. Each letter of the word *FRACTION* is written on a card and placed in a bag. What is the probability of choosing a *C*?

   a. How many possible outcomes are there? List them.

   _____ 8; F, R, A, C, T, I, O, N _____

   b. How many favorable outcomes are there? List them. ____1; C____

   c. You can write the probability of choosing a *C* as a fraction.

   $$P(C) = \frac{\text{number of favorable outcomes}}{\text{number of possible equally likely outcomes}}$$

   Write the probability as a fraction. ____$\frac{1}{8}$____

3. Each letter of the word *NUMBER* is written on a card and placed in a bag. What is the probability of choosing a vowel?

   a. How many possible outcomes are there? List them.

   _____ 6; N, U, M, B, E, R _____

   b. How many favorable outcomes are there? List them.

   _____ 2; U, E _____

   c. You can write the probability of choosing a vowel as a fraction.

   $$P(\text{vowel}) = \frac{\text{number of favorable outcomes}}{\text{number of possible equally likely outcomes}}$$

   Write the probability as a fraction in simplest form. ____$\frac{1}{3}$____

Reteach **RW93**

Name _____

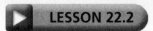

Problem Solving Skill: Too Much or Too Little Information

Some problems seem confusing because you are given more information than you need. One way to help understand the problem is to first circle the question asked. Second, underline only the information you need to answer the question. Third, solve the problem.

Read the following problem.

Christine tries calling 5 friends to ask about her homework. She knows that she must do problems 5–12, but she cannot remember the page number. She knows it is either page 150, 152, 154, or 156. Her friends aren't home, so she guesses. What is the probability that she will do the problems on the correct page?

Step 1   Circle the question asked.

You are asked to find the probability that she will do the problems on the correct page.

Step 2   Underline the information you need.

You know the homework is on one of these pages: 150, 152, 154, or 156.

You know she guesses.

Step 3   Solve the problem.

The probability that she will choose the correct page is $\frac{1}{4}$.

Write if each problem has too much, too little, or the right amount of information. Then solve the problem if possible, or write what information is needed to solve it.

1. Corey is waiting at a bus stop for the Number 3 bus. If all the buses that use the bus stop are equally likely to come along, what is the probability that the next bus will be the Number 3 bus?

   too little; need number of buses

2. Twelve boys and 13 girls answered a survey about their favorite food. Fifteen said pizza, 6 said French fries, and the rest said hamburgers. What is the probability that a student picked at random did not say pizza?

   too much; $\frac{2}{5}$

RW94   Reteach

Name _____

LESSON 22.4

Experimental Probability

You have a box containing thousands of radish seeds. How many of them will sprout? You can experiment to find out.

Suppose you plant 20 of the seeds, and 15 of them sprout.

$$\frac{15 \text{ seeds sprout}}{20 \text{ seeds in all}} \rightarrow \frac{15}{20} = \frac{3}{4}$$

You might conclude, based on your experiment, that about $\frac{3}{4}$ of the seeds will sprout. So, if you select a seed at random from the box, the **experimental probability** that it will sprout is $\frac{3}{4}$.

Mel spins a spinner numbered 1 to 5. The table below shows the results of 25 spins.

Number	1	2	3	4	5
Frequency	4	2	6	8	5

1. What is the experimental probability of spinning a 4?

   a. How many times did Mel spin a 4? ____8____

   b. How many times did Mel spin in all? ____25____

   c. You can write the experimental probability as a fraction.

   $$\text{Experimental probability} = \frac{\text{number of times outcome occurs}}{\text{total number of trials}}$$

   Use the formula to write the experimental probability of spinning 4 as a fraction in simplest form. ____$\frac{8}{25}$____

2. Suppose Mel spins the spinner 50 more times. How often can he expect to spin a 3?

   a. Begin by finding the experimental probability of spinning a 3.

   How many times did Mel spin a 3? ____6____

   b. How many times did Mel spin? ____25____

   c. Use the formula to write the experimental probability as a fraction in simplest form. ____$\frac{6}{25}$____

   d. Multiply 50 spins by the experimental probability.

   $$50 \times \frac{6}{25} = 12$$

   e. How many times can Mel expect to spin a 3? ____12 times____

Reteach RW95

Name _____

Problem Solving Strategy: Make an Organized List

When you read a problem, you need to decide on a problem solving strategy to solve it. One problem solving strategy that often works well when you want to find combinations of things is *Make an Organized List*.

Juan is planning a camping trip. He can go in June, July, or August to Yosemite, Yellowstone, or Grand Teton. How many different camping trips involving one place and one month are possible?

Step 1 Think about what you want to find and what facts are given.

- You are asked to find how many different camping trips involving one place and one month are possible.
- There are 3 months in which Juan can go and 3 different locations.

Step 2 Decide on a strategy for solving.

- Since you need to match months and locations in order to find the number of camping trips, you can use the strategy *Make an Organized List*.

Step 3 Solve.

- Follow through with the strategy. List the 3 months and pair each one with the three camping trips. Then count the number of choices in your list.

June/Yosemite	July/Yosemite	August/Yosemite
June/Yellowstone	July/Yellowstone	August/Yellowstone
June/Grand Teton	July/Grand Teton	August/Grand Teton

There are 9 possible choices for Juan's camping trip.

Solve the problem by making an organized list.

1. Craig and his friends want to go to a movie. There are 3 movies playing. They each begin at 12:30 P.M., 6:00 P.M., 8:30 P.M., and 10:30 P.M. How many different movie choices do Craig and his friends have?

   _____ 12 choices _____

2. Mr. Campo is taking a train to visit a friend. He can take a local train or an express train. They both leave the station at the same times: 11:30 A.M., 4:30 A.M., and 8:00 A.M. How many choices does Mr. Campo have?

   _____ 6 choices _____

3. Ms. Jackson sells sweat suits in her sporting goods store. They come in either a medium or large size. Both sizes come in red, navy, green, or black. How many different choices of a sweat suit does Ms. Jackson offer?

   _____ 8 choices _____

4. Liz wants to buy a bouquet of flowers. She has her choice of tulips, roses, or carnations. She can get each kind of flower in red, white, or yellow. She can have a fancy vase or a plain vase. How many choices does Liz have?

   _____ 18 choices _____

Name _____

LESSON 23.2

Compound Events

A compound event includes 2 or more simple events. One way to find the number of possible outcomes for a compound event is by using the **Fundamental Counting Principle:**

> To find the total number of possible outcomes, multiply the number of possible outcomes of each event.

Tiwa is buying a pizza. She has the choice of 2 sizes, 2 crusts, and 4 toppings. How many choices of size, crust, and topping does she have?

Event 1		Event 2		Event 3		
Sizes		Crusts		Toppings		
2	×	2	×	4	=	16

Tiwa has 16 choices.

Use the Fundamental Counting Principle to find the number of possible outcomes for each situation.

1. a choice of 3 salads and 4 dressings

   _____ 12 outcomes _____

2. a choice of 6 ice-cream flavors and 5 toppings

   _____ 30 outcomes _____

3. Mark's family is buying a new car. They want a station wagon or a sedan. The colors they like are black, white, and burgundy, with either a black or gray interior. How many choices do they have?

   _____ 12 choices _____

4. Carla wants to buy a sandwich. She can have either white or whole wheat bread. She has her choice of 4 fillings with either mayonnaise or mustard. How many sandwich choices does she have?

   _____ 16 choices _____

5. The Sideline Cafe offers 10 different sandwiches, 3 different salads, and 4 different soups. How many different sandwich-salad-soup combinations does the cafe offer?

   _____ 120 combinations _____

6. Miko is buying fabric to make curtains. She has a choice of 8 patterns in 12 colors. The fabric comes in cotton or rayon. How many fabric choices does she have?

   _____ 192 choices _____

Reteach RW97

Name _____

LESSON 23.3

Independent and Dependent Events

To determine if an event is an independent event or a dependent event, ask yourself if the first event affects the second event.

Independent Events	**Dependent Events**
If the outcome of the first event doesn't affect the outcome of the second event, the events are **independent** events.	If the outcome of the first event affects the outcome of the second event, the events are **dependent** events.
To find the probability of two independent events happening, multiply the probability of the first event by the probability of the second event.	To find the probability of two dependent events happening, multiply the probability of the first event by the probability of the second event.

If A and B are independent events, then $P(A, B) = P(A) \times P(B)$	If A and B are dependent events, then $P(A, B) = P(A) \times P(B \text{ after } A)$

A bag contains 6 lettered tiles labeled E, V, E, N, T, and S. Without looking, Emilio selects a tile and replaces it. Then he selects a second tile. What is the probability of Emilio selecting an E and then a T?

Find P(E, T)

first event: $P(E) = \frac{2}{6} = \frac{1}{3}$

second event: $P(T) = \frac{1}{6}$

$P(E, T) = \frac{1}{3} \times \frac{1}{6} = \frac{1}{18}$

The probability of Emilio selecting an E, replacing it, and then selecting a T is $\frac{1}{18}$.

A bag contains 6 lettered tiles labeled E, V, E, N, T, and S. Without looking, Emilio selects a tile and keeps it. Then he selects a second tile. What is the probability of Emilio selecting an E and then a T?

Find P(E, T)

first event: $P(E) = \frac{2}{6} = \frac{1}{3}$

second event: $P(T \text{ after } E) = \frac{1}{5}$

$P(E, T) = \frac{1}{3} \times \frac{1}{5} = \frac{1}{15}$

The probability of Emilio selecting an E, keeping it, and then selecting a T is $\frac{1}{15}$.

A jar contains cards with all the letters of the word "ALGEBRA." Without looking, you take a card out of the jar and replace it before selecting again. Find the probability of each event. Then find the probability assuming the card is not replaced after each selection.

1. P(L, R)

   $\frac{1}{49}, \frac{1}{42}$

2. P(A, E)

   $\frac{2}{49}, \frac{1}{21}$

3. P(A, A)

   $\frac{4}{49}, \frac{1}{21}$

4. P(G, A)

   $\frac{2}{49}, \frac{1}{21}$

5. P(R, not A)

   $\frac{5}{49}, \frac{2}{21}$

6. P(B, L or R)

   $\frac{2}{49}, \frac{1}{21}$

7. P(A, not R)

   $\frac{12}{49}, \frac{5}{21}$

8. P(R, vowel)

   $\frac{3}{49}, \frac{1}{14}$

RW98 Reteach

Name _____

LESSON 23.4

Make Predictions

A population is a group being studied. A sample is a representative part of that group. You can use the results of a sample to make predictions about a population.

A random sample of 100 sixth graders showed that 20 of them choose green as their favorite color. What is the probability that a randomly selected sixth grader chooses green as his or her favorite color?

Step 1 Write a ratio: $\dfrac{20 \text{ students prefer green}}{100 \text{ students in sample}}$

Step 2 Write in simplest form: $\dfrac{20}{100} = \dfrac{2}{10} = \dfrac{1}{5}$

The probability that a sixth grader chooses green as his or her favorite color is about $\dfrac{1}{5}$.

There are 240 sixth graders at Lincoln Middle School. Predict about how many of them would choose green as their favorite color.

Step 1 Use the ratio to write a proportion: $\dfrac{1}{5} = \dfrac{n}{240}$

Step 2 Find the cross products: $1 \times 240 = 5 \times n$
$240 = 5n$

Step 3 Solve the equation: $\dfrac{240}{5} = \dfrac{5n}{5}$
$48 = n$

So, about 48 sixth graders would choose green as their favorite color.

Use Data For 1–3, use the table. The table shows the favorite music indicated by a random sample of 100 sixth graders from Thorton Middle School.

FAVORITE MUSIC	
Music	Number of Students
pop	40
classical	15
rock	25
country	20

1. If there are 280 sixth graders at Thorton Middle School, about how many would prefer pop music?

   _____ about 112 students _____

2. If there are 420 sixth graders at Thorton Middle School, about how many would prefer a kind of music other than pop music?

   _____ about 252 students _____

3. If there are 250 sixth graders at Thorton Middle School, about how many would prefer either pop music or country music?

   _____ about 150 students _____

Reteach RW99

Name _____

LESSON 24.1

Algebra: Customary Measurements

Ricki needs to buy 90 in. of wire for a science project. The store sells the wire by the foot. How many feet of wire does Ricki need?

Divide to change smaller units to larger units.

90 in. = ■ ft

Think: 12 in. = 1 ft

Divide 90 by 12.

So, 90 in. = $7\frac{1}{2}$ ft.

$$7\frac{6}{12}, \text{ or } 7\frac{1}{2}$$
$$12\overline{)90}$$
$$\underline{-84}$$
$$6$$

Ricki needs $7\frac{1}{2}$ ft of wire.

Juan has 9 qt of punch. He wants to give each classmate 1 c. How many classmates can he serve?

Multiply to change larger units to smaller units.

9 qt = ■ c

Think: 1 qt = 4 c

Multiply 9 by 4. $9 \times 4 = 36$

So, 9 qt = 36 c.

Juan can serve 36 classmates.

Solve.

1. Bill bought $8\frac{1}{4}$ yd of rope. How many inches of rope did Bill buy?

   _____297 in._____

2. Rosa has 8 gal of juice. How many quarts of juice does she have?

   _____32 qt_____

3. Terri has $10\frac{1}{2}$ yd of fabric. How many inches of fabric does she have?

   _____378 in._____

4. Jim's backpack weighs 88 oz. How many pounds does the backpack weigh?

   _____$5\frac{1}{2}$ lb_____

5. Gary has 44 ft of hedges along his fence. How many yards of hedges does he have?

   _____$14\frac{2}{3}$ yd_____

6. Monica has a water pitcher that holds 48 oz of water. How many quarts of water does the pitcher hold?

   _____$1\frac{1}{2}$ qt_____

7. Lana poured 24 cups of lemonade. How many pints did she pour?

   _____12 pints_____

8. Which is heavier, $2\frac{1}{2}$ tons or 5,000 pounds?

   _____They are equal._____

Reteach

Name _____

LESSON 24.2

Algebra: Metric Measurements

Mel made 1.45 L of soup. How many milliliters of soup does he have?
You can write a proportion to change metric units of measurement.

Step 1
Use the relationship of liter to milliliters to write a proportion.

$$\frac{1 \text{ L}}{1{,}000 \text{ mL}} = \frac{1.45 \text{ L}}{x \text{ mL}}$$

Step 2
Find the cross products.

$$x = 1{,}000 \times 1.45$$

Step 3
Solve for x.

$$x = 1{,}450$$

So, Mel has 1,450 mL of soup.

Use a proportion to solve. Check students' proportions.

1. Barbara has 2.7 km of ribbon. How many meters of ribbon does she have?

    _____2,700 m_____

2. A box weighs 8.7 kg. How many grams does it weigh?

    _____8,700 g_____

3. Rico has 536 m of rope. How many kilometers of rope does he have?

    _____0.536 km_____

4. The chef made 2,300 mL of gravy. How many liters of gravy did he make?

    _____2.3 L_____

5. Vince has 19.5 L of paint. How many mL of paint does he have?

    _____19,500 mL_____

6. Beth's suitcase weighs 1.9 kg. How many grams does it weigh?

    _____1,900 g_____

7. Rachel has made 3,500 mL of punch. How many liters of punch does she have?

    _____3.5 L_____

8. Akeem has 15.3 kg of books. How many grams of books does he have?

    _____15,300 g_____

9. A package weighs 12.3 kg. How many grams does it weigh?

    _____12,300 g_____

10. Mary's new car is 3 m long. How many kilometers long is her new car?

    _____0.003 km_____

Reteach RW101

Name _____

LESSON 24.3

Algebra: Relate Customary and Metric

Two systems of measurement are used in the United States, customary and metric. A length, weight, or capacity might be expressed in one of these systems when you need to know the equivalent measure in the other system. You need to know the relationships between the two systems to find equivalent measures.

A recipe for biscuits uses 250 g of flour. You would like to make the biscuits but need to know an equivalent customary measure, such as ounces. How many ounces of flour does the recipe require?

You must find the number of ounces that are equivalent to 250 g.

Step 1: Find the relationship between ounces and grams.
   1 oz ≈ 28.35 ← The symbol ≈ is read "is approximately equal to."

Step 2: Think: If 1 oz ≈ 28.35 g, then 2 oz ≈ 2 × 28.35, or 56.7 g.
   Continue in this way until you reach a value close to 250 g.

Ounces	1	2	3	4	5	6	7	8	9
Grams	28.35	56.7	85.05	113.4	141.75	170.1	198.45	226.8	255.15

From the table, you can see that 250 g is approximately equal to 9 oz.

Another method uses the relationship between ounces and grams to write a proportion.

$$\frac{1 \text{ oz}}{28.35 \text{ g}} = \frac{x \text{ oz}}{250 \text{ g}}$$

You can then solve the proportion for x to see how many ounces are approximately equivalent to 250 g.

$$28.35x = 250 \quad \text{Find the cross products.}$$
$$x = \frac{250}{28.35}$$
$$x \approx 8.82$$

So, you will need about 9 oz of flour for the biscuit recipe.

Use a proportion to convert to the given unit.

1. 1 kg ≈ 2.2 lb
   7 kg ≈ __?__ lb

   15.4

2. 1 m ≈ 3.28 ft
   3 m ≈ __?__ ft

   9.84

3. 1 qt ≈ 0.95 L
   9 L ≈ __?__ qt

   9.47

4. 1 in. ≈ 2.54 cm
   65 in. ≈ __?__ cm

   165.1

5. 1 mi ≈ 1.61 km
   9 km ≈ __?__ mi

   5.59

6. 1 yd ≈ 0.91 m
   250 yd ≈ __?__ m

   227.5

RW102 Reteach

Name _____

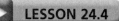

Appropriate Tools and Units

You can measure the length of the same line segment using different scales on a ruler. The smaller the unit you use, the more precise the measurement will be.

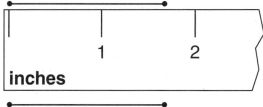 Measured to the nearest inch, the line segment is 2 in. long.

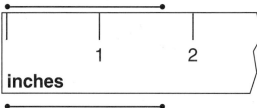 Measured to the nearest half inch, the line segment is $1\frac{1}{2}$ in. long.

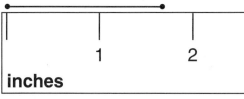 Measured to the nearest quarter inch, the line segment is $1\frac{3}{4}$ in. long.

Notice in the example that there was a different result each time the segment was measured. The result found by measuring to the nearest quarter inch, $1\frac{3}{4}$ in., is the most precise measurement of the segment.

Measure the line segment to the given length.

1. nearest inch; nearest half inch

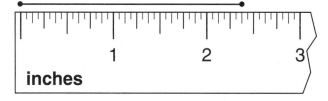

2 in.; $2\frac{1}{2}$ in.

2. nearest centimeter; nearest millimeter

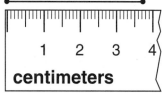

4 cm; 37 mm

3. nearest half inch; nearest quarter inch

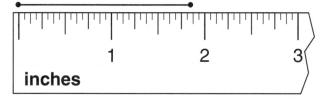

2 in.; $1\frac{3}{4}$ in.

4. nearest centimeter; nearest millimeter

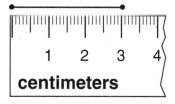

2 cm; 23 mm

5. nearest centimeter; nearest millimeter

5 cm; 48 mm

6. nearest half inch; nearest quarter inch

1 in.; $1\frac{1}{4}$ in.

Reteach **RW103**

Name _____

Problem Solving Skill: Estimate or Find Exact Answer

When you read a problem, you should find an exact answer only if you have decided that an estimate is not sufficient. However, many problems that you will encounter can be solved by estimating.

Questions you can ask to help you decide on the type of answer include:

- Is it possible to find an exact answer?
  You may only be able to estimate when numbers such as large populations are involved.
- Will finding an exact answer help me in any way?
  You may need, for example, to know an exact amount of money or an exact count of people.
- Is an estimate enough to solve the problem?
  Many problems ask you only to decide whether a certain minimum amount has been reached.

A mail-order company has these 8 orders to ship. The shipping company charges $3.95 per package if the average weight of all the packages is less than 3 lb. Does the mail-order company qualify for the $3.95 rate?

| shirts: 4 lb | ties: 1 lb | shoes: 4 lb | shirts: 2 lb |
| shirts: 2 lb | jeans: 3 lb | shirts: 2 lb | gloves: 1 lb |

Step 1: Think about what you know and what you are asked to find.
- You know the weight of each package.
- You are asked to find whether the average weight of the 8 packages is less than 3 lb.

Step 2: Decide on a plan to solve.
- Since you need to find only whether the average weight of the packages is less than 3 lb, you do not need an exact answer. Estimate the average weight.

Step 3: Solve.
- Carry out the plan. Note that there are only 2 packages that weigh more than 3 lb. Both packages weigh just 1 lb more than the 3 lb average. There are 5 packages that each weigh less than 3 lb. So, it is safe to say that the average weight of the packages is less than 3 lb.

The mail-order company qualifies for the $3.95 shipping rate.

Tell whether an estimate or an exact answer is needed. Solve.

1. How much will the mail-order company have to pay the shipper to ship the 8 packages?

   exact; $31.60

2. On a day when 200 packages are to be shipped at the $3.95 rate, will the shipping cost be greater than $1,000?

   estimate; no

Name _____

LESSON 25.2

Perimeter

The **perimeter** of a polygon is the distance around it. To find the perimeter, add the lengths of the sides of the polygon.

Rick wants to find the perimeter of this polygon.
Find the unknown length. Then find the perimeter.

Look at the base of the figure. It has a length of 24 m.
The length of the top is equal to 24m. The figure shows that two of the pieces that make up this length are each 6 m. So, $x = 24 - 12$, or 12 m.

Add the lengths of all the sides.

$12 + 6 + 6 + 12 + 24 + 12 + 6 + 6 = 84$

So, the perimeter of the figure is 84 m.

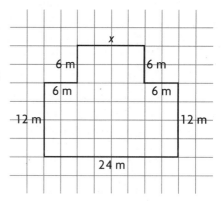

Find the unknown length. Then find the perimeter.

1.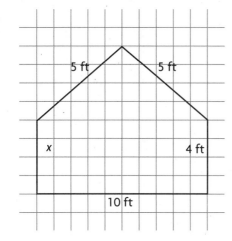

   $x = 4$ ft; 28 ft

2.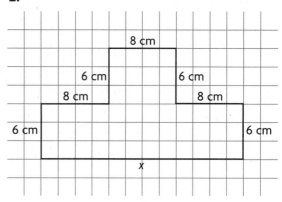

   $x = 24$ cm; 72 cm

3.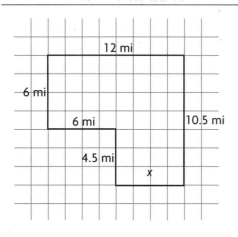

   $x = 6$ mi; 45 mi

4.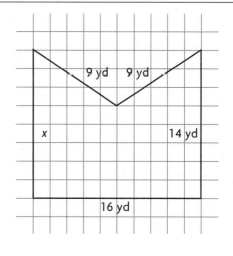

   $x = 14$ yd; 62 yd

Reteach RW105

Name _____

LESSON 25.3

Problem Solving Strategy: Draw a Diagram

Kerry made a rectangular picture frame from 4 wood strips. The perimeter was 72 in. The length was 8 in. greater than the width. The next frame she made was as wide as the first but was 8 in. longer. It had a perimeter of 88 in. Find the dimensions of each frame.

Step 1: Think about what you know and what you are asked to find.

- You know the following about the first frame:
  It was rectangular in shape.
  The perimeter was 72 in.
  The length was 8 in. greater than the width.

- You know the following about the second frame:
  The width was the same as the first frame.
  The length was 8 in. longer than the length of the first frame.
  The perimeter was 88 in.

- You are asked to find the dimensions (length and width) of each frame.

Step 2: Plan a strategy to solve.

- Use the strategy *draw a diagram*.
- Draw the first frame using information in the problem. Experiment until you find a rectangle that meets the requirements.
- Draw the second frame from what you have learned from the first diagram.

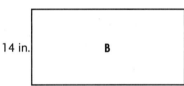

Step 3: Solve.

- Carry out the strategy.
  From the problem, you learn that the only rectangle that meets the requirements has length 22 in. and width 14 in. From the first diagram, you learn that the dimensions of the second frame must be length 30 in. and width 14 in.

Use the strategy *draw a diagram* to help you solve.

1. Kerry made a triangular frame. Two sides were the same length. The third was 6 in. longer than the other two. The perimeter of the frame was 33 in. How long was each side?

   _____ 9 in., 9 in., 15 in. _____

2. Two square frames required a total of 144 in. of wood. The perimeter of one frame was 16 in. greater than the perimeter of the other. What were the perimeters of each frame?

   _____ 64 in., 80 in. _____

RW106 Reteach

Name _____

LESSON 25.4

Circumference

If you understand the relationship between the diameter of a circle and its circumference, choosing the correct formula will be easier.

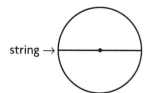
string →

- The string is stretched across the circle so that it touches the center point. The length of the string is equal to the diameter of the circle.

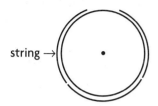
string →

- The string can be used to show the distance around, or circumference of the circle.
- It takes 3 string lengths plus a little more to equal the circumference. The extra that is needed is about $\frac{14}{100}$ or 0.14 of the string length.

The ratio of 3.14 diameters to 1 circumference is called *pi*. The symbol used to show this relationship is the Greek letter π. The relationship applies to all circles. If the length of a diameter or radius is given as a mixed number, it may be more convenient to use $\frac{22}{7}$ rather than 3.14.

So, the formula for the circumference of a circle is $C = \pi d$. Read the formula as "Circumference equals pi times the length of a diameter."

Since a diameter is equal to two radii (the plural of radius), the formula can also be written $C = 2\pi r$ or "Circumference equals two times pi times the length of a radius."

Find the circumference of the circle. Use 3.14 or $\frac{22}{7}$ for π.

1.
9 cm

2.
16 m

3.
$3\frac{1}{2}$ in.

_____ _____ _____

4.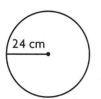
24 cm

5.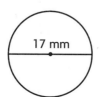
17 mm

6.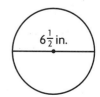
$6\frac{1}{2}$ in.

_____ _____ _____

Reteach RW107

Name _____

LESSON 26.1

Estimate and Find Area

You can find the area of a figure on a grid by counting the square units inside the figure.

- This is a full square.
- These squares are half full.
- These squares are more than half full.
  They are almost-full squares.
- These squares are less than half full.
  They are almost-empty squares.

Here is how to estimate the area of the figure at the right. Each small square on the grid represents $1\ ft^2$.

- full squares: Count 9. Include all 9.
- almost-full squares: Count 4. Include all 4.
- half-full squares: Count 4. Include $\frac{1}{2} \times 4$, or 2.
- almost-empty squares: Count 3. Do not include them.

Find the sum of the squares counted. $9 + 4 + 2 = 15$
The area of the figure is about $15\ ft^2$.

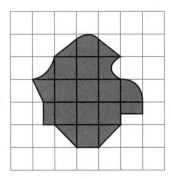

Complete to estimate the area of each figure. Each small square of the grid represents $1\ yd^2$. *Possible estimates are given.*

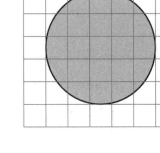

1. There are __13__ full squares.

   There are __8__ almost-full squares.

   There are __0__ half-full squares.

   There are __4__ almost-empty squares.

   __13__ + __8__ + ($\frac{1}{2}$ × __0__) = __21__   The area is about __21__ yd^2.

Estimate the area of each figure. Each small square on the grid represents $1\ ft^2$.
Possible estimates are given.

2.

3.

about $21\ ft^2$ about $18\ ft^2$

RW108 Reteach

Name _____

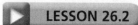

Algebra: Areas of Parallelograms and Trapezoids

This trapezoid is formed by a right triangle and a rectangle. Its area is the sum of the areas of the triangle and the rectangle.

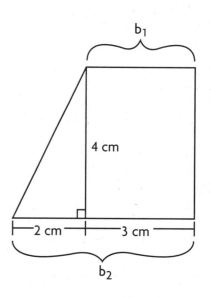

area of the triangle   area of rectangle

$A = \frac{1}{2}bh = 4 \text{cm}^2$ $A = bh = 12 \text{cm}^2$

Area of trapezoid = 4 cm² + 12 cm² or 16 cm²

Check by using the formula: $A = \frac{1}{2}h(b_1 + b_2)$

$A = \frac{1}{2} \times 4(3 + 5)$

$A = \frac{1}{2} \times 4 \times 8 = 16$

A right triangle of the same size but upside down is added to the opposite side of the rectangle to form this parallelogram. Its area is the sum of the areas of both triangles and the area of the rectangle.

triangle   triangle   rectangle   parallelogram

4cm² + 4cm² + 12cm² = 20cm²

The area of the parallelogram is 20 cm².

Check by using the formula: $A = bh$.

$A = 5 \times 4 = 20$

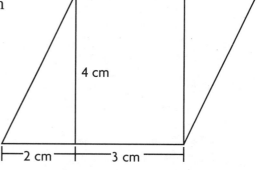

Find the area of each parallelogram. Use the formula $A = bh$.

1. $b = 15$ cm; $h = 8$ cm   2. $b = 7$ yd; $h = 5$ yd   3. $b = 9$ ft; $h = 7$ ft

   $A =$ __120__ cm² $A =$ __35__ yd² $A =$ __63__ ft²

Find the area of each trapezoid. Use the formula $A = \frac{1}{2}h(b_1 + b_2)$.

4. $b_1 = 5$ cm; $b_2 = 6$ cm   5. $b_1 = 12$ ft; $b_2 = 5$ ft   6. $b_1 = 3$ yd; $b_2 = 7$ yd
   $h = 8$ cm $h = 9$ ft $h = 5$ yd

   $A =$ __44__ cm² $A =$ __76.5__ ft² $A =$ __25__ yd²

Name _____

LESSON 26.4

Algebra: Areas of Circles

A circle has a radius (*r*) and a diameter (*d*). Here is how they are related.

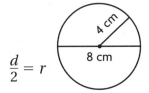

$d = 8$ cm $\qquad r = 4$ cm $\qquad 2 \times r = d \qquad \dfrac{d}{2} = r$

You can use the formula $A = \pi r^2$ to find the area of the circle.

$\pi \approx 3.14$

The symbol $\approx$ means "is approximately equal to."

$A \approx 3.14 \times 4^2$

Multiply. $\quad 3.14 \times 4 \times 4 = 50.24$

The area of the circle is about 50 cm².

Write the letter of the correct formula from Column 2.

            Column 1   Column 2

1. rectangle ____b____   a. $A = \pi r^2$

2. parallelogram ____d____   b. $A = lw$

3. circle ____a____   c. $A = \dfrac{1}{2} bh$

4. triangle ____c____   d. $A = bh$

Use the formula $A = \pi r^2$ to find the area of each circle. Round to the nearest whole number. All areas are approximations.

5.   6.   7.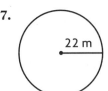

    __201 in.²__   __907 cm²__   __1,520 m²__

8. $r = 9$ yd __254 yd²__   9. $r = 5$ ft __79 ft²__   10. $r = 7$ in. __154 in.²__

11. $r = 10$ m __314 m²__   12. $d = 12$ cm __113 cm²__   13. $d = 30$ in. __707 in.²__

14. $d = 100$ mm __7,850 mm²__   15. $r = 25$ yd __1,963 yd²__   16. $d = 9$ m __64 m²__

RW110   Reteach

Name _____

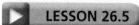

Algebra: Surface Areas of Prisms and Pyramids

Tony wants to wrap a present. He needs to know how much wrapping paper to buy. The box is shaped like a rectangular prism. How can you find the number of square inches of wrapping paper are needed to wrap the present?

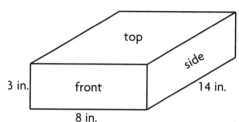

You need to find the area of each of the 6 faces of the box, or the surface area of the box. Each face is a rectangle, so use the formula for the area of a rectangle.

The opposite faces have the same area.

Find the area of the front. Find the area of the top. Find the area of one side.
$A = lw$ $A = lw$ $A = lw$
$\quad = 8 \times 3 = 24$ $\quad = 14 \times 8 = 112$ $\quad = 14 \times 3 = 42$

Then multiply each area by 2 to include the opposite faces.

front and back: top and bottom: left and right sides:
$24 \times 2 = 48$ $112 \times 2 = 224$ $42 \times 2 = 84$

Then add the areas. $48 + 224 + 84 = 356$

So, 356 in.^{2} of wrapping paper is needed to wrap the present.

Find the surface area for each rectangular prism.

1.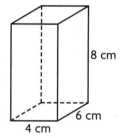

   208 cm^2

2.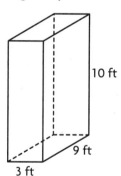

   294 ft^2

3.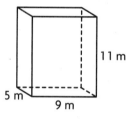

   398 m^2

4.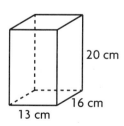

   1,576 cm^2

5.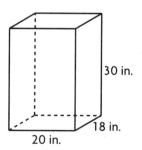

   3,000 in.^{2}

6.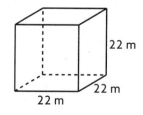

   2,904 m^2

Reteach RW111

Name _____

LESSON 27.1

Estimate and Find Volume

Hilary has two boxes—a rectangular box that measures 3 ft by 4 ft by 5 ft and a triangular box that measures 3 ft by 4 ft in length and width and 5 ft in height. What is the volume of each box?

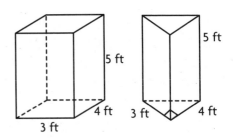

You can find the volume by using the formula $V = Bh$.

V stands for volume, B stands for the area of the base, and h stands for the height.

For a rectangular prism, any side can be the base, since they are all rectangles. For a triangular prism, the base is the triangular side.

Find the volume of the rectangular box.

Step 1 Find the area of the base.

The base is 3 ft by 4 ft.
$A = l \times w$
$A = 3 \text{ ft} \times 4 \text{ ft} = 12 \text{ ft}^2$

Step 2 Multiply the area of the base by the height.

$V = Bh$
$12 \text{ ft}^2 \times 5 \text{ ft} = 60 \text{ ft}^3$

So, the volume of the rectangular box is 60 ft³.

Find the volume of the triangular box.

Step 1 Find the area of the base.

The base is a right triangle.
$A = \frac{1}{2} bh$
$A = \frac{1}{2} \times 3 \text{ ft} \times 4 \text{ ft} = 6 \text{ ft}^2$

Step 2 Multiply the area of the base by the height.

$V = Bh$
$6 \text{ ft}^2 \times 5 \text{ ft} = 30 \text{ ft}^3$

So, the volume of the triangular box is 30 ft³.

Find the volume of each figure.

1.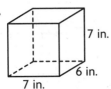

2.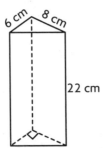

3.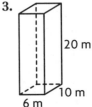

  294 in.³   528 cm³   1,200 m³

RW112 Reteach

Name _____

LESSON 27.2

Problem Solving Strategy: Make a Model

How does doubling the dimensions of a rectangular prism affect the volume?

You can answer this question by comparing the models of two rectangular prisms.

Rectangular Prism 1

The rectangular prism below has dimensions of 2 cm × 3 cm × 4 cm.

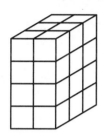

Rectangular Prism 2

The rectangular prism below has dimensions of 4 cm × 6 cm × 8 cm.

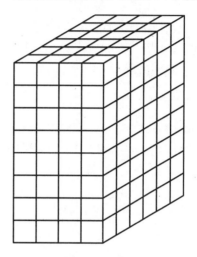

The volume is
2 cm × 3 cm × 4 cm = 24 cm³.

The volume is
4 cm × 6 cm × 8 cm = 192 cm³.

Compare the volume of Prism 1 with the volume of Prism 2.

192 ÷ 24 = 8

So, the volume of Prism 2 is 8 times as great as the volume of Prism 1.

Find the volume of each prism. Then double the dimensions and find the volume.

	Length	Width	Height	Volume	Volume (Doubled Dimensions)
1.	5 m	2 m	6 m	60 m³	480 m³
2.	3 ft	2 ft	8 ft	48 ft³	384 ft³
3.	5 in.	6 in.	7 in.	210 in.³	1,680 in.³
4.	10 cm	10 cm	10 cm	1,000 cm³	8,000 cm³
5.	8 m	15 m	2 m	240 m³	1,920 m³
6.	12 cm	14 cm	15 cm	2,520 cm³	20,160 cm³

Reteach RW113

Name _____

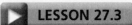

Algebra: Volumes of Pyramids

To find the volume of a pyramid use the formula below.

Words	Volume	=	$\tfrac{1}{3}$	times	area of base	times	height of figure
⇓	↓		↓	↓	↓	↓	↓
To	Volume (V)	=	$\tfrac{1}{3}$	×	Base (B)	×	height (h)
⇓							
Formula	V	=	$\tfrac{1}{3}Bh$				

Step 1 Find the area of the base. This pyramid has a rectangular base. Use $l \times w$ = area.

$l \times w$ = area of base

$7 \times 8 = 56$ in.^{2}

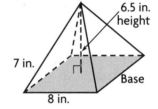

Step 2 Use the area of the base to find the volume. Multiply base times height.

$V = \tfrac{1}{3}Bh$

$V = \tfrac{1}{3} \times 56 \times 6.5 = \tfrac{1}{3} \times 364$

Step 3 Then multiply the product of B and h by $\tfrac{1}{3}$.

$V = \tfrac{1}{3} \times 364$

$V = \tfrac{364}{3}$

$V = 121\tfrac{1}{3}$

The volume of the pyramid is $121\tfrac{1}{3}$ in.^{3}

Find the volume.

1.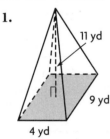

   132 yd^3

2.

   160 m^3

3.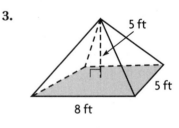

   $66\tfrac{2}{3}$ ft^3

4. rectangular pyramid:
   l = 7 in., w = 4.5 in., h = 8 in.

   84 in.^{3}

5. square pyramid:
   l = 13 ft, w = 13 ft, h = 15 ft

   845 ft^3

Name _____

LESSON 27.5

Volumes of Cylinders

The cylinder at the right has a height of 8 in. and a radius of 2 in. A cylinder has a circular base.

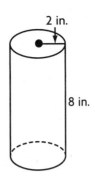

You can find the volume of a cylinder by using the formula $V = Bh$.
The B stands for the area of the base, which is a circle.

Step 1 Find the area of the base.
The base is a circle.

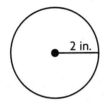

Use the formula $A = \pi r^2$.
Use 3.14 for π.
$A \approx 3.14 \times 2 \text{ in.} \times 2 \text{ in.} \approx 12.56 \text{ in.}^2$

Step 2 Multiply the area of the base, B, times the height.

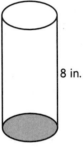

$12.56 \text{ in.}^2 \times 8 \text{ in.} \approx 100.48 \text{ in.}^3$

So, the volume of the cylinder is about 100 in.³ to the nearest whole number.

Find the volume of each cylinder. Round to the nearest whole number.

1.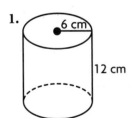

   about 1,356 cm³

2.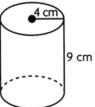

   about 452 cm³

3.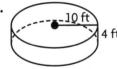

   about 1,256 ft³

4.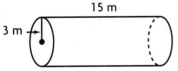

   about 424 m³

5.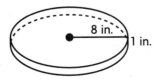

   about 201 in.³

6.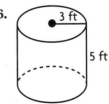

   about 141 ft³

Reteach RW115

Name _____

LESSON 28.1

Problem Solving Strategy: Find a Pattern

A problem solving strategy that often works well when events repeat is *find a pattern*.

Ms. Jacobs owns a bookstore. She gets a discount on the number of books she orders. If she orders 4 copies of a book, she will get a $2 discount. For 8 copies, she will get a $5 discount; for 12 copies, an $8 discount; for 16 copies, an $11 discount; and so on until she reaches the maximum discount of $20. How many books does Ms. Jacobs need to order to get the maximum discount?

Step 1 Think about what you want to find and what facts are given.
- You want to find how many books Ms. Jacobs needs to order to get the maximum discount of $20.
- You are given the amount of the discount for the number of books that are ordered.

Step 2 Decide on a strategy for solving.
- Since the number of books increases by the same amount, and the discount increases by the same amount you can use the strategy *Find a Pattern*.

Step 3 Solve.
- Find the patterns:

    books ordered: 4, 8, 12, 16 → increases by 4 each time
    discount: $2, $5, $8, $11 → increases by $3 each time

- Show the patterns in a table to find the number of books that must be ordered to get the maximum discount of $20.

Number of Books	4	8	12	16	20	24	28
Amount of Discount	$2	$5	$8	$11	$14	$17	$20

Ms. Jacobs must order 28 books to get the maximum discount of $20.

Solve the problem by finding a pattern.

1. Alberto and Toshi are on a 7-day hike. They hike 10 mi the first day; 12 mi the second day; 14 mi the third day; and 16 mi the fourth day. If this pattern continues, how many miles will they hike on the seventh day?

    _____ 22 miles _____

2. Janell has decided to jog 1 mi each day this week. She plans to jog $1\frac{1}{2}$ mi each day next week, 2 mi each day the next week, and so on until she jogs 5 mi each day. If this pattern continues, when will that happen?

    _____ the ninth week _____

RW116 Reteach

LESSON 28.2

Name _____

Patterns in Sequences

John rides the bus to work every day. The bus fare each day is $3.25. On payday John budgets $20.00 for transportation. How much of this money will John have after 5 workdays?

You can use a pattern to solve this problem.

Start: $20.00
After 1 day: $20.00 − $3.25 = $16.75
After 2 days: $16.75 − $3.25 = $13.50
After 3 days: $13.50 − $3.25 = $10.25
After 4 days: $10.25 − $3.25 = $7.00
After 5 days: $7.00 − $3.25 = $3.75

Subtract $3.25 from each term to find the next term.

If you write the terms as a sequence, you have
$20.00, $16.75, $13.50, $10.25, $7.00, $3.75, ...

So, after 5 days John will have $3.75.

Write a sequence and solve.

1. Jeff is starting an exercise program. The first week he will jog 5 mi. If he increases the number of miles by 2 mi each week, in how many weeks will he be jogging more than 14 mi?

   _____ 5, 7, 9, . . . ; 6 weeks _____

2. Pat spends $2.45 for lunch each day. She has $20.00 budgeted for lunch. On which day will she need additional lunch money?

   _____ $20.00, $17.55, $15.10, . . . ; day 9 _____

3. Danielle is reading a book. The first day she reads 3 pages. If she increases the number of pages she reads by 3 pages each day, on which day will she be reading more than 20 pages?

   _____ 3, 6, 9, . . . ; day 7 _____

4. During a five-week reading program, Brad read 2 books the first week. If he increases the number of books he reads by 2 each week, how many books will he read the fifth week?

   _____ 2, 4, 6, . . . ; 10 books _____

Reteach

Name _____

LESSON 28.3

Number Patterns and Functions

Very often, one quantity depends on another quantity. You can write a function to show that relationship.

Eric wants to mail a letter that weighs 6 ounces. The cost of mailing the letter is $.033 for each ounce or part of an ounce. How much will it cost Eric to mail his letter?

Step 1 Look for a pattern:

weight	cost
1 ounce	$0.33
2 ounces	$0.66
3 ounces	$0.99
4 ounces	$1.32

The pattern shows that as the number of ounces increases by one, the cost increases by $0.33.

Step 2 Use the pattern to write an equation.
Let w = number of ounces and c = total cost.
$c = 0.33 \times w$

Step 3 Solve the equation.
$c = 0.33 \times 6$
$c = \$1.98$

It will cost Eric $1.98 to mail his letter.

Solve the problem by finding a pattern.

1. Carolyn wants to mail a package that weights 5 lb. The cost of mailing the package is $0.75 for each pound or part of a pound. Write an equation and solve to find out how much it will cost Carolyn to mail her package.

   _____ $c = \$0.75w; \$3.75$ _____

3. Nina is making a picture frame that is 24 in. long. She wants the length to be three times the width. Write an equation and solve to find out how wide she should make the frame.

   _____ $w = l \div 3; 8$ in. _____

3. Jack wants to buy a video player for $150.00. He earns $6.25 per hr doing yard work. Write an equation and solve to find out how many hours Jack will have to work to earn enough money to buy the video player.

   _____ $h = m \div \$6.25; 24$ hr _____

4. All 32 members of Sam's computer club are going on a field trip. Each van can hold 6 passengers besides the driver. Write an equation and solve to find out how many vans they will need.

   _____ $v = p \div 6; 6$ vans _____

Reteach

Name _____

LESSON 28.4

Geometric Patterns

There are many types of patterns based on geometric figures. Study these patterns.

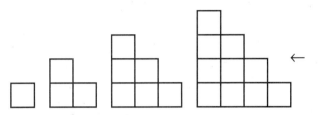

1 square, 3 squares, 6 squares, 10 squares, . . .
The next figure will have 15 squares.

The pattern above is based on the arrangement of the squares. The number of squares on the left side and on the bottom row increases by 1. So the next figure will be 5 squares high and 5 squares wide.

This pattern is based on the type of shape.

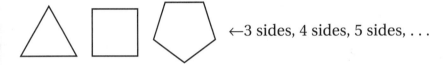

←3 sides, 4 sides, 5 sides, . . .

So, the next figure in the pattern will have 6 sides.

A pattern can be based on shading, such as the following pattern. The figure doesn't change, except for the shading pattern.

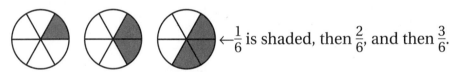

← $\frac{1}{6}$ is shaded, then $\frac{2}{6}$, and then $\frac{3}{6}$.

So, the next figure in the pattern will have $\frac{4}{6}$ shaded.

Draw the next two figures in the geometric pattern.

1.

2.

Reteach RW119

Name _____

LESSON 29.1

Transformations of Plane Figures

A **transformation** is a movement that doesn't change the size or shape of a figure.

Three kinds of transformations are:

Translation	Rotation	Reflection
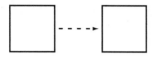		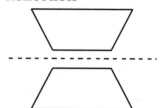
Slide the figure along a straight line.	Turn the figure around a point.	Flip the figure over a line.

Tell whether the drawing shows a translation, rotation, or reflection.

1.
2.
3.

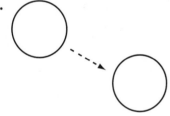 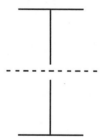 

____translation____   ____reflection____   ____rotation____

4.
5.
6.

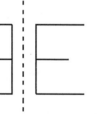 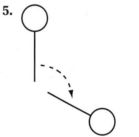 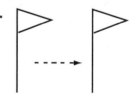

____reflection____   ____rotation____   ____translation____

7.
8.
9.

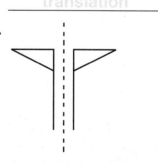

____rotation____   ____translation____   ____reflection____

RW120 Reteach

LESSON 29.2

Name _____

Tessellations

A tessellation is an arrangement of shapes that completely covers a flat surface.

The simplest tessellations involve squares or rectangles. You have seen examples on kitchen floors or on walls, such as the following.

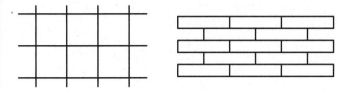

To make a more interesting pattern, start with any square or rectangle.

   ←Cut out a piece of any shape from one side.

   ←Translate the piece to the opposite side. Then tape it.

Now tessellate the new shape. Here are two ways to do so.

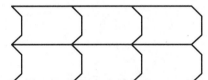   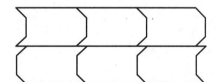

Draw the tessellation shape described by each pattern. Then form two rows of a tessellation. Check students' tessellations.

1.

2.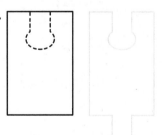

Reteach RW121

Name _____

LESSON 29.3

Problem Solving Strategy: Make a Model

Larry is designing a tile mosaic for a floor. The shape he uses must tessellate a plane. He wants to use the shape at the right. Can he use this shape for his design?

To solve this problem, you can make a model.

Trace and cut out several of the shapes Larry wants to use. Use the cutouts to see if the shape will tessellate a plane.

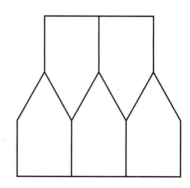

Move the shapes around to see if they will fit together without gaps or overlaps.

Check that the sum of the angle measures around a point is 360°.

The shape Larry wants to use will tessellate the plane.

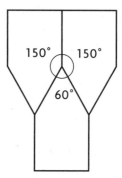

Make a model to solve.

1. Erika is making a design from the shape below. She wants the design to tessellate. Can she use this shape?

   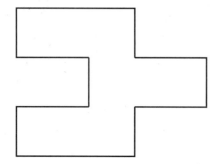

   _____ yes

2. For a tabletop design, Shawn uses octagons and squares. Will his design tessellate a plane?

   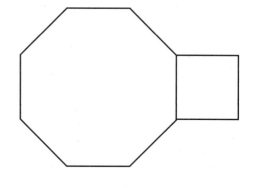

   _____ yes

RW122   Reteach

Name _____

LESSON 29.4

Transformations of Solid Figures

Find the number of unique ways you can place this figure on top of the gray plane figure. Like a puzzle, the prism must fit over the gray square exactly.

Step 1 Look and think about what you are asked to do. The figure is a prism. It has two parallel faces that are squares. It also has two sets of parallel faces that are rectangles. The gray plane figure is a two-dimensional square.

Step 2 Find the faces of the prism that will cover the gray square. The rectangular faces do not fit. So find every way each square face will fit over the gray square.

Step 3 The prism in the position of the original figure counts as the first way. Rotate around an imaginary vertical axis to find the first 4 ways.

Step 4 Next, rotate the prism around a horizontal axis so that you find each way the other square face can be placed.
Again, rotate the prism around a vertical axis to find the final 4 ways it will fit. There are a total of 8 ways that the prism fits.

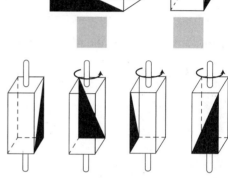
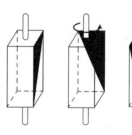

Write the number of ways you can place the solid figure on the plane figure in the second column and on the plane figure in the third column.

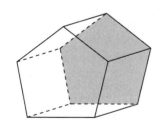

1.

2.

_____ 10 ways _____ _____ 10 ways _____

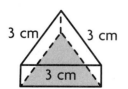

3.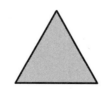

4.

_____ 6 ways _____ _____ 6 ways _____

Reteach **RW123**

Name _____

LESSON 29.5

Symmetry

A figure has **line symmetry** if it can be folded so that the two parts of the figure match.

Look at the top figure.

The dashed line is a line of symmetry.

If you fold the figure on the dashed line, the two parts match.

The bottom figure shows another line of symmetry.

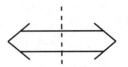
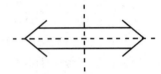

Draw the line or lines of symmetry for each figure.

1.
2.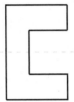
3.
4.

5.
6.
7.
8.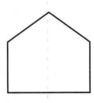

Draw a figure that is not shown above that has at least two lines of symmetry. Check students' drawings.

Name _____

LESSON 30.1

Inequalities on a Number Line

An inequality uses the symbols <, >, ≤, or ≥. You can use a number line to show the solution of an algebraic inequality.

Graph all of the solutions to $b > 7$.

Step 1
Draw a number line that has the number 7 on it and at least two numbers to the left and right of 7.

Step 2
Test the number 7 by substituting it into the inequality.
 $b > 7;\ 7 > 7$ False
So, the graph does not include the number 7.

Step 3
Decide which side of 7 is the solution. Do this by picking a number to the left of 7 and a number to the right of 7.
To the left of 7 is 6, and to the right of 7 is 8.

Step 4
Test each number. Graph points for the solutions to the side that is true.
 Left: $b > 7;\ 6 > 7$ False
 Right: $b > 7;\ 8 > 7$ True
All values greater than 7 are included in the solution. The circle around 7 means that 7 is not included in the solution.

For each inequality, test three numbers: the number itself, the whole number to the left, and the whole number to the right. Write *true* or *false*.

1. $x > 13$
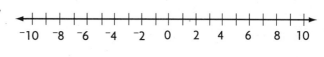

2. $x \leq 9$

3. $x \geq {}^{-}2$

Graph each inequality on a number line.

4. $x < 5$

5. $x \leq {}^{-}6$
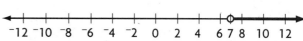

Reteach RW125

Name _____

LESSON 30.2

Graph on the Coordinate Plane

Finding coordinates or graphing points on a coordinate plane can be done using these steps.

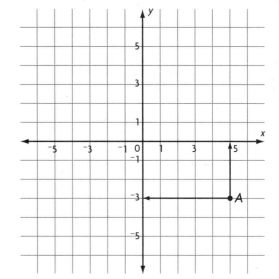

Finding Coordinates

To find the coordinates for point A:
- From the point, move straight up to the x-axis.
- Read the number. 5 is the x-coordinate.
- From the point, move to the left to the y-axis.
- Read the number. ⁻3 is the y-coordinate. The ordered pair is (5,⁻3).

Graphing Points

To graph point B (⁻2,⁻4):
- The first number indicates direction left or right. Starting at the origin, move 2 units to the left on the x-axis. Draw a vertical line through this point.
- The second number indicates direction up or down. Starting at the origin, move 4 units down. Draw a horizontal line through this point.
- The point where the vertical line and horizontal line cross is the point B (⁻2,⁻4).

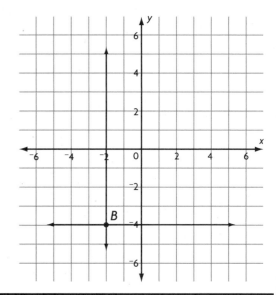

Write the ordered pair for each point on the coordinate plane.

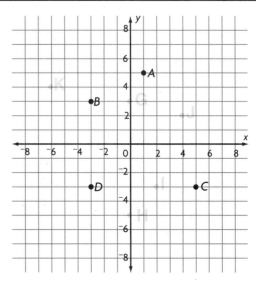

1. point A
   _____(1,5)_____

2. point B
   _____(⁻3,3)_____

3. point C
   _____(5,⁻3)_____

4. point D
   _____(⁻3,⁻3)_____

Graph each point on the coordinate plane.

5. G (0,3)
6. H (0,⁻5)
7. I (2,⁻3)
8. J (4,2)
9. K (⁻6,4)

RW126 Reteach

Name _____

LESSON 30.3

Graph Functions

This is a function table.
The first row represents the
values of x, and the second
row represents the values
of y. To graph the function,
you need to

x	⁻3	⁻2	⁻1	0	1	2	3
y	3	4	5	6	7	8	9

- write ordered pairs like (⁻3, 3).
- locate the ordered pairs on a coordinate plane.

Ordered Pairs

(⁻3,3); (⁻2,4); (⁻1,5); (0,6); (1,7); (2,8); (3,9)

Graph Ordered Pairs

You can also write an algebraic equation
by looking at the pattern in the table.

- The value of y is 6 more than the value of x.

$$y = x + 6$$

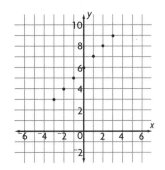

A

x	⁻3	⁻2	⁻1	0	1
y	1	2	3	4	5

B

x	⁻3	⁻2	⁻1	0	1
y	9	6	3	0	⁻3

1. List the ordered pairs for Table A.

   (⁻3,1); (⁻2,2); (⁻1,3); (0,4); (1,5)

2. List the ordered pairs for Table B.

   (⁻3,9); (⁻2,6); (⁻1,3); (0,0); (1,⁻3)

3. Locate the points for the ordered pairs of the function from Exercise 1.

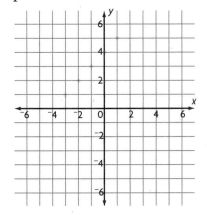

4. Locate the points for the ordered pairs of the function from Exercise 2.

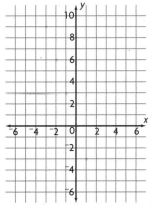

5. Write an equation relating y to x using the data in Table A.

   $y = x + 4$

6. Write an equation relating y to x using the data in Table B.

   $y = {-3}x$

Reteach RW127

Name _____

LESSON 30.4

Problem Solving Skill: Make Generalizations

One problem solving strategy that often works well when one quantity depends on another is to make a generalization. You can show a generalization by writing an equation that describes the relationship between the two quantities.

Each of the tables in the school cafeteria can seat 8 students. How many tables does the cafeteria need to seat 128 students?

Step 1 Use the information you are given to make a table. Let x stand for one quantity and y for the other quantity. Then look for a pattern.

Number of Tables (x)	1	2	3	4	5
Number of Students (y)	8	16	24	32	40

The table shows that each x-value is $\frac{1}{8}$ the corresponding y-value.

Step 2 Use the pattern to write an equation.

Number of Tables ⟶ $x = \frac{1}{8}y$ ⟵ Number of Students

Note that the equation can be used to find the number of tables (x) for any number of students (y). That's why the equation is called a generalization.

Step 3 Solve the equation.
To solve the equation, replace y with the number of students.

$y = 128$ $x = \frac{1}{8} \times 128$ $x = \frac{128}{8}$ $x = 16$

The cafeteria needs 16 tables to seat 128 students.

Write an equation to show the generalization. Then solve.

1. Each of the tables in the school cafeteria can seat 12 students. How many tables (x) will the cafeteria need to seat 252 students (y)?

    _____ $x = \frac{1}{12}y$; 21 tables

2. Mark is riding on a train pulled by an old steam engine. The train averages a speed of 45 mi per hr. How many miles (y) will Mark travel in 8 hr (x)?

    _____ $y = 45x$; 360 mi

RW128 Reteach

Name _____

LESSON 30.5

Graph Transformations

The three basic transformations for figures in a coordinate plane are translations, reflections, and rotations.

- A translation moves the figure right or left, or up or down, or both.

  Figure ABC is translated up and right. Follow point A as it moves 4 units up and 3 units right. The coordinates of A' are (1,4). The new figure is triangle A'B'C'.

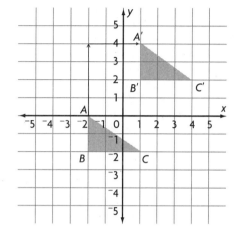

- A reflection is a flip across a line. Rectangle DEFG is reflected across the y-axis.

  After a point is reflected, the distance from the line stays the same. Both E and E' are 1 unit from the y-axis.

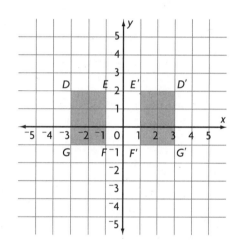

- A rotation about the origin can be clockwise or counterclockwise. It can be any number of degrees.

  Think of tracing paper placed over the graph and attached at the origin. You can draw the figure on the tracing paper and then rotate it.

  Triangle HJK is rotated 90° clockwise about the origin.

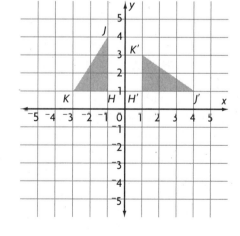

The coordinates of triangle PQR are (0,0), (2,0), and (0,4). Find the new coordinates after the transformation of triangle PQR.

1. a translation 4 units to the right

   (4,0), (6,0), and (4,4)

2. a reflection across the x-axis

   (0,0), (2,0), and (0,⁻4)

3. a 90° counterclockwise rotation about the origin

   (0,0), (0,2), and (⁻4,0)

Reteach RW129